Analog Circuits and Signal Processing

Series editors

Mohammed Ismail, Dublin, USA
Mohamad Sawan, Montreal, Canada

More information about this series at http://www.springer.com/series/7381

Yu Lin · Hans Hegt · Kostas Doris
Arthur H.M. van Roermund

Power-Efficient High-Speed Parallel-Sampling ADCs for Broadband Multi-carrier Systems

Yu Lin
NXP Semiconductors
Eindhoven
The Netherlands

Kostas Doris
NXP Semiconductors
Eindhoven
The Netherlands

Hans Hegt
Department of Electrical Engineering
Eindhoven University of Technology
Eindhoven
The Netherlands

Arthur H.M. van Roermund
Department of Electrical Engineering
Eindhoven University of Technology
Eindhoven
The Netherlands

ISSN 1872-082X ISSN 2197-1854 (electronic)
Analog Circuits and Signal Processing
ISBN 978-3-319-36891-7 ISBN 978-3-319-17680-2 (eBook)
DOI 10.1007/978-3-319-17680-2

Springer Cham Heidelberg New York Dordrecht London

Softcover reprint of the hardcover 1st edition 2015

Printed on acid-free paper

Springer International Publishing AG Switzerland is part of Springer Science+Business Media
(www.springer.com)

Contents

Symbols and Abbreviations

A	Scaling factor
ADC	Analog-to-digital converter
ADE	Cadence analog design environment
ADSL	Asymmetric digital subscriber line
AMP	Amplifier
AWG	Arbitrary waveform generator
BER	Bit error rate
C	Capacitance (F)
CICC	IEEE Custom Integrated Circuits Conference
CMOS	Complementary metal oxide semiconductor
CR	Clipping ratio
CS	Current source
CS-DAC	Current steering digital-to-analog converter
$\mathrm{D}_{\mathrm{aux}}$	Auxiliary path data output
$\mathrm{D}_{\mathrm{out}}$	Digital data output
$D(n)$	Digital data symbol
DAC	Digtial-to-analog converter
DFT	Discrete Fourier transform
DNL	Differential non-linearity (LSB)
DOCSIS	Data over cable service interface specification
DUT	Device under test
$E\{\cdot\}$	Expectation
ECG	Electrocardiography
ENOB	Effective number of bits (bit)
ERBW	Effective resolution bandwidth (Hz)
ESSCIRC	IEEE European Solid State Circuits Conference
$f(x)$	Probability density function
$F(x)$	Distribution function
f_c	Carrier frequency
f_{cs}	Subcarrier spacing
FE	Front end

FFT	Fast Fourier transform
FinFET	Fin-shaped field-effect transistor
FOM	Figure of merit
FS	Full scale amplitude measured by a single sinusoid (V)
f_s	Sampling rate (Hz)
$g(t)$	Pulse function
GSM	Global system for mobile communications
GS/s	Gigahertz sample per second
HD	Harmonic distortion
HVQFN	Heatsink very-thin quad flat-pack no-leads
I/O	Input/output
I^2C	Inter-integrated circuit
IC	Integrated circuit
IEEE	Institute of Electrical and Electronics Engineers
IFFT	Inverse fast Fourier transform
IMD	Inter-modulation distortion
INL	Integral nonlinearity (LSB)
ISSCC	IEEE international Solid-State Circuits Conference
k	Boltzmann's constant (Joules/Kelvin)
L_{clip}	Clipping level of an ADC (V)
LSB	Least significant bit
LTE	Long-Term Evolution
MC-CDMA	Multi-carrier code division multiple access
MDAC	Multiplying digital-to-analog converter
MUX	Multiplexer
NPR	Noise power ratio
os	Offset voltage (V)
OFDM	Orthogonal frequency division multiplexing transmission
PAPR	Peak-to-average power ratio
$P_{backoff}$	Power back-off value (dB)
P_c	Clipping distortion power (W)
PCB	Printed circuit board
Pdf	Probability density function
P_{fs}	ADC full scale signal power measured with a sinusoid (W)
P_n	Noise power (W)
P_s	Signal power (W)
PSD	Power spectral density (W/Hz)
$Q(x)$	Quantization function
QADC	Quarter ADC
QAM	Quadrature amplitude modulation
R_{ADC}	Signal range processed by the ADC
RMS	Root mean square
SAR	Successive approximation register
SC	Switched capacitor
SCDR	Signal to clipping distortion ratio

SelMUX	Path selection multiplexer
SFDR	Spurious free dynamic range
SHA	Sample-and-hold-amplifier
SNCDR	Signal to noise and clipping distortion ratio
SNDR	Signal to noise and distortion ratio
SNR	Signal-to-noise ratio
SoC	System on chip
SOI	Silicon on insulator
SQNR	Signal to quantization noise ratio
STNR	Signal to thermal noise ratio
T/H	Track and hold
THD	Total harmonic distortion
TI-ADC	Time interleaving ADC
T_b	Data block duration (sec)
T_s	Sampling period (sec)
USB	Universal serial bus
V_{aux}	Auxiliary path input signal (V)
V_b	Bias voltage (V)
V_{clk}	Clock voltage (V)
V_{cm}	Common mode voltage (V)
V_{dd}	Supply voltage (V)
V_{in}	Input voltage (V)
V_{main}	Main path input signal (V)
$V_{n.RMS}$	RMS noise voltage (V)
V_{out}	Output voltage (V)
V_{ppd}	Peak-to-peak differential voltage (V)
V_r	Reference voltage (V)
V_{res}	Output residue signal voltage (V)
$V_{s.RMS}$	RMS signal voltage (V)
V_{th}	Threshold voltage (V)
VLSI	Symposia on VLSI Technology and Circuits
WiMAX	Worldwide interoperability for microwave access
WLAN	Wireless local area network
WPAN	Wireless personal area network
Δt	Timing skew (sec)
Δ_{LSB}	ADC least significant bit size (V)
η_{vol}	Voltage efficiency
η_{vol}	Current efficiency
Φ	Clock signal

Chapter 1
Introduction

1.1 Background

Electronic devices are pervasive in our daily life improving our life quality in almost every aspect: from connectivity to safety, from healthcare to entertainment and in many other aspects. The core of modern electronics is the integrated circuit (IC) which is an electronic circuit formed on a small piece of semiconducting material that performs the same function as a larger circuit made from discrete components [1]. Since its invention by Jack Kilby and Robert Noyce in the 1950s [2] and with the continuous advances in process technology, ICs empower electronic devices with ever increasing digital signal processing power and capacity to store information digitally. However, real-world signals, such as electromagnetic waves, sound, motion, pressure and temperature, are continuously variable physical quantities due to the nature of the world. In order to exploit the digital signal processing power and data storage capability of ICs, data converters including analog-to-digital converters (ADCs) and digital-to-analog converters (DACs) are essential building blocks in modern electronic systems that bridge the gap between the analog and digital "worlds". The number of applications that require data converters, including sensing applications, process control and instrumentation, digital audio and video applications, health care and life sciences, wire/wireless communication terminals and infrastructure, satellite communication and military communication applications and so on, is extremely large and constantly expanding.

Despite the fact that data converters have already gone a long way of development since their inventions, more advanced data converters are still demanded in terms of better accuracy, higher speed, lower power consumption and smaller die size. Major driving forces behind this demand nowadays are: the trend of ever increasing user demands for higher data throughput in the field of communication and the trend of system-on-chip (SoC) integration and shifting more and more signal processing functions from the analog domain into the digital domain for

Y. Lin et al., *Power-Efficient High-Speed Parallel-Sampling ADCs for Broadband Multi-carrier Systems*, Analog Circuits and Signal Processing,
DOI 10.1007/978-3-319-17680-2_1

lower cost and higher flexibility [3]. Data converters are mixed-signal ICs which have both digital and analog circuits. The digital circuits benefit significantly from the scaling of the complementary metal oxide semiconductor (CMOS) technology. The cost per logic function is constantly reducing as the density of digital logics doubles every 2–3 years over the last few decades [4], and operation speed of digital logics and their power consumption are also improving with the on-going technology scaling. However, analog circuit design becomes more complicated in order to meet the increasing performance demands and challenges associated with the technology scaling. In advanced CMOS technology, performance of analog circuits is negatively affected due to the reduction of the transistor's intrinsic gain, the increase in gate and subthreshold leakage currents, as well as the reduction of supply voltage and signal swing, and so on [5]. Therefore, the continuous demands for more advanced data converters and challenges of analog circuit design in advanced CMOS technology make data converters always an active research topic in both academia and industry.

1.2 Book Aim and Outline

To address the above mentioned demands and challenges, the "smart data converters" concept which implies context awareness, on-chip intelligence and adaptation was proposed [6–9]. The core of this concept is in exploiting various information either a priori or a posteriori (obtained from devices, signals, applications or the ambient situations, etc.) for circuit and architecture optimization during design phase or adaptation during operation to enhance data converters performance/flexibility/robustness/power-efficiency and so on. Many works, including those from the Mixed-signal Microelectronics group of TU/e [10–14], have contributed to the development of this concept and demonstrated various methods and techniques to enable smartness of data converters.

The aim of this book is to contribute to the development and application of the "smart data converters" concept. The main focus lies on exploiting the a priori knowledge of the system/application to develop enhancement techniques for ADCs and with a particular emphasis on improving the power efficiency of high-speed and high-resolution ADCs for broadband multi-carrier systems. In contrast to the conventional ADC design approach, a dedicated ADC architecture called "parallel-sampling architecture" is developed based on previous works [15–17] by making use of the multi-carrier signal properties (a priori knowledge of the system) to tailor the ADC's analog core circuitry without compromising system performance. This architecture has been applied to high-speed and high-resolution pipeline and time-interleaving SAR ADCs for broadband multi-carrier communication systems. The validation of the concept was carried out with IC implementations and demonstrated state-of-the-art power efficiency compared to ADCs with a similar performance for broadband multi-carrier signals.

The outline of this book is as follows:

Chapter 2 gives a short introduction to the basics of Nyquist-rate ADCs and discusses the performance limitations and trends of ADCs. Then, a system a priori knowledge aware design concept is presented. Various ADC design examples that exploit special properties of the signals for performance enhancement are reviewed.

Chapter 3 presents a parallel-sampling technique to enhance the power efficiency of ADCs for broadband multi-carrier systems based on the concept introduced in Chap. 2. Firstly, multi-carrier transmission is introduced, the multi-carrier signal statistics properties are analyzed and their impacts on the ADC dynamic range requirement are investigated. Secondly, power efficient design techniques for high-speed thermal noise limited ADCs are reviewed. Finally, a parallel-sampling ADC architecture for multi-carrier systems is presented and analyzed in detail. By exploiting the statistical properties of multi-carrier signals, this technique can be applied to ADCs for converting a larger input signal without causing excessive clipping distortion and with improvement in accuracy over the critical small amplitude region. Therefore, a better overall signal to noise and clipping distortion ratio can be achieved without using a conventional higher resolution ADC. This technique allows reducing power dissipation and area in comparison to conventional approach of using larger devices to lower thermal noise for converting multi-carrier signals.

Chapter 4 is devoted to the application of the parallel-sampling technique to the conventional pipeline and time-interleaving SAR ADCs. In this chapter, architecture studies and circuit implementations of a parallel-sampling first stage for a 200 MS/s 12 b switched-capacitor pipeline ADC using 65 nm CMOS technology and a parallel-sampling frontend stage for a 4 GS/s 11 b time-interleaved ADC using 40 nm CMOS technology were presented; circuit simulations are shown and discussed. Furthermore, a design example as a validation of the concept introduced in Chaps. 2 and 3 was presented. The implementation details of a parallel-sampling ADC IC which contains a dual 11 b 1 GS/s time-interleaved SAR ADC is described. The IC is implemented in 65 nm CMOS technology and tested to verify the basic concept that is described in Chap. 3. Furthermore, the experimental results are compared with prior art with a similar performance.

Finally, Chap. 5 concludes this book and provides prospects for future work based on the insight gained during this research work.

References

1. *Oxford English Dictionary*, 3rd Ed. Oxford University Press, 2010. [online].
2. *The Nobel Prize in Physics 2000*. Nobelprize.org. Nobel Media AB 2013. http://www.nobelprize.org/nobel_prizes/physics/laureates/2000/. Accessed 12 May 2014.
3. Kester, W.A. 2005. *Data Converters History*. Data conversion handbook. Amsterdam: Elsevier.
4. Brock, D. 2006. *Understanding Moore's law: Four decades of innovation*. Philadelphia, Pa: Chemical Heritage Foundation.

5. Manganaro, G. 2012. *Advanced data converters*. Cambridge: Cambridge University Press.
6. van Roermund, A.H.M., Hegt, J., Harpe, P., Radulov, G., Zanikopoulos, A., Doris, K., and Quinn, P.: Smart AD and DA converters, in *IEEE International Symposium on Circuits and Systems*, 2005, pp. 4062–4065.
7. van Roermund, A.H.M.: Smart, flexible, and future-proof data converters. In *18th European Conference on Circuit Theory and Design*, 2007, pp. 308–319.
8. van Roermund, A.H.M. An integral Shannon-based view on smart front-ends. In *European Conference on Wireless Technology*, 2008, pp. 25–28.
9. van Roermund, A.H.M.: Shifting the Frontiers of Analog and Mixed-Signal Electronics" (submitted for publication).
10. Doris, K., A.H. van Roermund, and D. Leenaerts. 2006. *Wide-bandwidth high dynamic range D/A converters*, 2006th ed. Mahwah, N.J.: Springer.
11. Harpe, P., H. Hegt, and A.H.M. van Roermund. 2012. *Smart AD and DA conversion, 2010*. Dordrecht: Springer.
12. Radulov, G., P. Quinn, H. Hegt, and A.H.M. van Roermund. 2011. *Smart and flexible digital-to-analog converters 2011*. Dordrecht: Springer.
13. van Beek, P.C.W. 2011. *Multi-carrier single-DAC transmitter approach applied to digital cable television*, Ph.D. dissertation, Technische Universiteit Eindhoven.
14. Tang,Y., Hegt, H., and van Roermund. A. 2013. *Dynamic-mismatch mapping for digitally-assisted DACs*, 2013 edition. New York:Springer.
15. Gregers-Hansen, V., Brockett, S.M., and Cahill, P.E. 2011. A stacked A-to-D converter for increased radar signal processor dynamic range. In *Proceedings of the 2001 IEEE Radar Conference, 2001*, pp. 169–174.
16. Doris, K. 2008. Data processing device comprising adc unit, WO2008072130 A1, 19 June 2008.
17. Cruz, P., and Carvalho, N.B. 2010. Architecture for dynamic range extension of analogue-to-digital conversion. In *2010 IEEE International Microwave Workshop Series on RF Front-ends for Software Defined and Cognitive Radio Solutions (IMWS)*, 2010, pp. 1–4.

Chapter 2
Enhancing ADC Performance by Exploiting Signal Properties

Abstract This chapter starts with a brief introduction of the analog-to-digital conversion process in Sect. 2.1 and a discussion of factors that define the performance of ADCs in Sect. 2.2. ADC performance limitations and trends are addressed in Sect. 2.3. In Sect. 2.4, a brief discussion of popular Nyquist-rate ADC topologies is given where the topologies most relevant to the focus of this book are discussed with the associated tradeoffs. A signal/system-aware design approach which exploits certain signal properties to enhance the ADC performance is discussed in Sect. 2.5 and examples are shown.

2.1 Introduction to Analog-to-Digital Converters

An analog-to-digital converter is an electronic circuit which converts a continuous-time and continuous-amplitude analog signal to a discrete-time and discrete-amplitude signal [1]. The analog-to-digital conversion involves three functions, namely sampling, quantizing and encoding [2], as shown in Fig. 2.1. After the conversion, the continuous quantities have been transformed into discrete quantities with a certain amount of error due to the finite resolution of the ADC and imperfections of electronic components. The purpose of the conversion is to enable digital processing on the digitized signal.

ADCs are essential building blocks in electronic systems where analog signals have to be processed, stored, or transported in digital form. The ADC can be a stand-alone general purpose IC, or a subsystem embedded in a complex system-on-chip (SoC) IC. A main driving force behind the development of ADCs over the years has been the field of digital communications due to continuous demand of higher data rates and lower cost [2]. In Fig. 2.2, a block diagram of a typical digital communication system is shown and the location of the ADC in the system is indicated [3]. The ADC is normally preceded by signal conditioning blocks (e.g. amplifiers, filters, mixers, modulators/demodulators, detectors, etc.) and followed by the baseband digital signal processing unit. With the advance in CMOS process

Y. Lin et al., *Power-Efficient High-Speed Parallel-Sampling ADCs for Broadband Multi-carrier Systems*, Analog Circuits and Signal Processing,
DOI 10.1007/978-3-319-17680-2_2

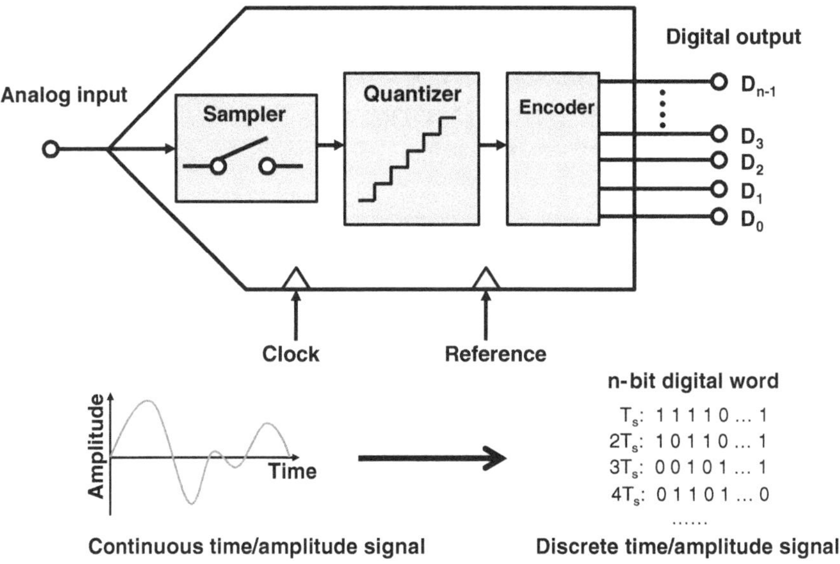

Fig. 2.1 Block diagram of an analog-to-digital converter

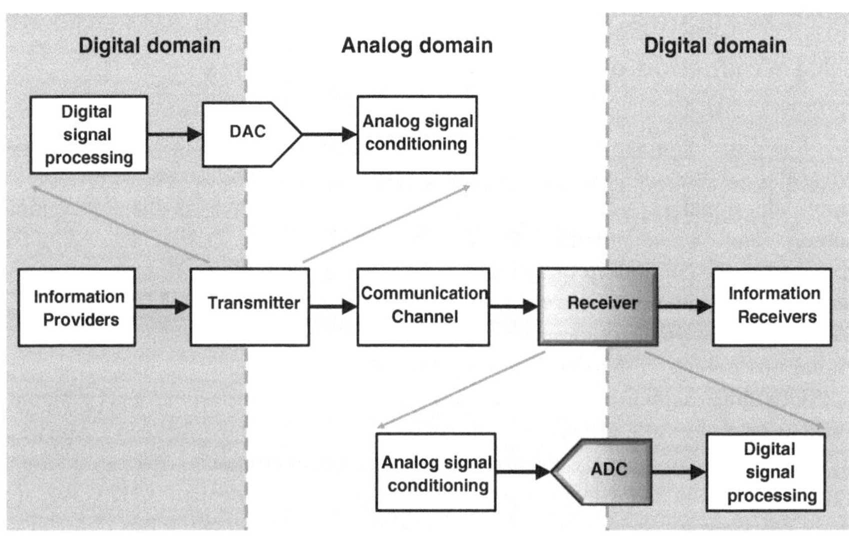

Fig. 2.2 Location of an ADC in a digital communication system

technology, the cost per digital function goes down exponentially. More and more signal conditioning functions are shifted from the analog processing domain into the digital processing domain e.g. to save cost or improve flexibility of the system [4, 5]. Data converters (ADCs and DACs) become crucial building blocks and even bottlenecks in a digital communication system [6]. Improvements of the ADC performance such as sampling rate, accuracy, and power consumption enable new system architectures and define the competitiveness of the overall solution.

2.2 ADC Performance Parameters

Depending on the context and applications, requirements for the ADC vary dramatically. Many parameters are used to define the performance of an ADC [1, 2, 7]. The purpose of using these parameters is to characterize the physical behavior of an ADC in order to specify, design, and verify it for targeted applications. This section reviews three key ADC parameters for digital communication systems which are conversion accuracy, bandwidth, and power.

2.2.1 Conversion Accuracy

The conversion accuracy refers to the degree of closeness of the ADC's output value to its actual input value and can be expressed in absolute or relative terms [2]. Ideally, the conversion accuracy is only limited by the ADC's references, the number of quantization levels and their spacing which decides how small the conversion error can be. In reality, the conversion error is always larger due to physical imperfections of electronic components which introduce noise and distortion to the signal. An abstract model of an ADC with typically encountered error sources is drawn in Fig. 2.3 to show what affects the conversion accuracy.

The degradation of conversion accuracy due to these errors can be quantified by static and dynamic performance parameters [1, 2].

The static performance of an ADC is typically quantified by offset error, gain error, the differential non-linearity (DNL) error and integral non-linearity (INL) error [2]. The DNL is defined as the difference after gain and offset correction between the actual step width and the ideal value of one least significant bit (LSB). The INL is defined as the deviations of the values on the actual transfer function from a straight line. The DNL and INL errors are caused by component mismatch due to fabrication process variations, mechanical stress, temperature gradients across the circuit and operation conditions.

The dynamic performance of an ADC is normally quantified by signal-to-noise ratio (SNR), signal-to-noise-and-distortion ratio (SNDR), effective number of bits (ENOB), total harmonic distortion (THD), spurious free dynamic range (SFDR), inter-modulation distortion (IMD), and noise power ratio (NPR) [7]. Degradation of the dynamic performance of an ADC is contributed not only by static errors, but also by noise and signal dependent non-idealities, such as thermal noise, clock jitter, power supply noise, cross-talk, comparator metastability, dynamic settling

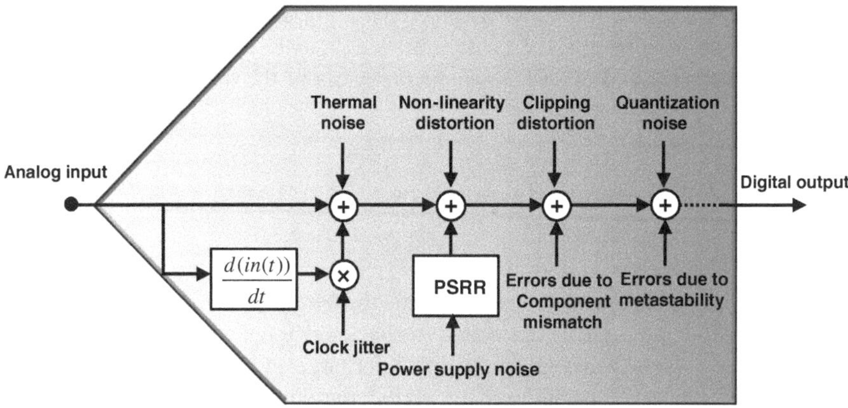

Fig. 2.3 An abstract model of an ADC with typical error sources

and non-linear transfer functions of internal circuit blocks, and so on. Depending on the actual implementation, one of them can be the dominant error source.

Among the above mentioned specifications, SNDR is one of the widely used specifications for comparing the conversion accuracy among different ADCs as all noise and distortion components that affect the conversion accuracy are included [7]. The SNDR is defined as:

$$SNDR = \frac{P_{signal}}{\underbrace{P_{distortion}}_{\text{Distortion}} + \underbrace{P_{quantization\_noise} + P_{thermal\_noise} + P_{jitter\_noise} + P_{other\_noise}}_{\text{Total noise power}}} \tag{2.1}$$

where P_{signal} is the average input signal power, $P_{distortion}$ is the total distortion power, and $P_{quantization}$, $P_{thermal}$, P_{jitter} and P_{other} are quantization, thermal, jitter and other noise power respectively. In most of the publications, the SNDR is measured using a single sinusoidal signal with full scale power as an excitation. The SNDR depends on both the amplitude and frequency of the signal since some of the error sources such as nonlinear distortion and clock jitter are input signal dependent, as shown in the Eq. 2.1.

In Eq. 2.1, noise and nonlinear distortion show equal contribution to the value of the SNDR. However, they can have very different impact on the performance of a specific system. Some systems are more sensitive to the nonlinear distortion, such as radar and GSM base station receivers; while some systems are more sensitive to noise, such as spread spectrum receivers. In these systems, specifying the conversion accuracy of the ADC separately with the SNR and SFDR is more appropriate than with the SNDR.

For communication systems adopting broadband multi-channel or multi-carrier transmission techniques, such as MC-CDMA, LTE, WiMAX, ADSL, and broadband cable modem, the actual signal that the ADC processes has very different properties compared to that of a single sinusoid. Using a simple sinusoid signal as an excitation to characterize the conversion accuracy of an ADC does not give an accurate representation of the real-world condition in these applications. For such systems, NPR testing provides an accurate measure of the noise and distortion performance of an ADC in a more realistic condition of a broadband system [2, 7, 8]. Instead of using a single sinusoid signal, a test signal, comprised of band-limited flat Gaussian noise to the frequency range of interest and with a narrow band (channel) of the noise deleted by a notch filters or other means, is used as an excitation for the NPR testing. The NPR is proven to be a more appropriate performance parameter and has gained popularity in characterizing broadband systems [7, 8]. Figure 2.4a shows an example of an NPR test signal in the frequency domain. The NPR is defined by the ratio of signal power measured in a certain frequency band to the combined noise and distortion power measured inside the notched frequency band (both frequency bands having equal bandwidth), as illustrated in Fig. 2.4a. The noise and distortion power measured inside the notched frequency band reveals the amount of noise and distortion caused by the ADC to the notched frequency band. In case the power spectral density of the signal is flat, it gives the same value as the ratio

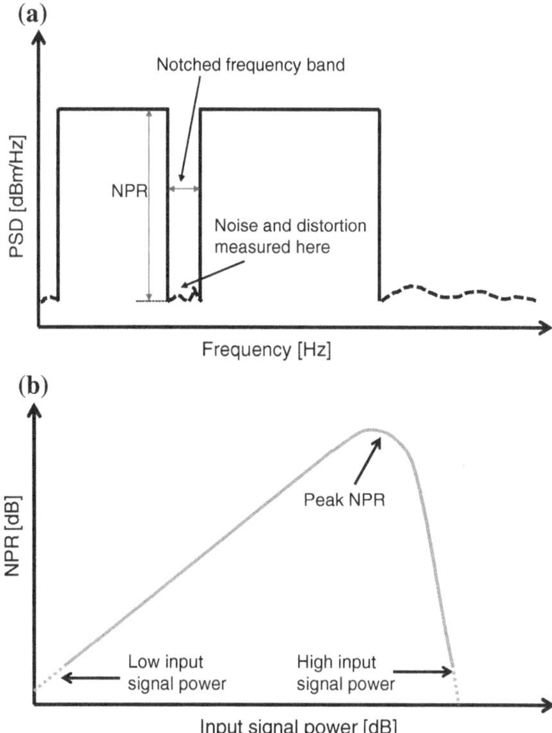

Fig. 2.4 **a** An example of a NPR test signal in the frequency domain; **b** NPR as a function of the test signal power

of the average power spectral density of the signal outside the notched frequency band to the average power spectral density inside the notched band as it is defined in [7]. The NPR is measured at the output of the ADC as the test signal is swept across a power range. Figure 2.4b shows a plot of NPR as a function of the test signal power.

The NPR is calculated, in decibels, from:

$$NPR = 10log_{10}\left(\frac{P_{No}}{P_{Ni}}\bigg|_{in\ equal\ BW}\right)dB = 10log_{10}\left(\frac{PSD_{No}}{PSD_{Ni}}\right)dB \tag{2.2}$$

where P_{No} and P_{Ni} are the power measured outside and inside the notched frequency band respectively, and PSD_{Ni} and PSD_{No} are the average power spectral density inside and outside the notched band respectively [7].

2.2.2 Bandwidth

Three commonly used definitions of the ADC bandwidth are the Nyquist bandwidth, the analog input bandwidth, and the effective resolution bandwidth (ERBW) [7]. The Nyquist bandwidth equals half of the sampling rate of the ADC. The sample rate (f_s) is the frequency at which the ADC converts the analog input waveform to digital data. The Nyquist theorem explains the relationship between the sample rate and the frequency content of the measured signal [9, 10]. The input signal bandwidth must be smaller than the Nyquist bandwidth to avoid aliasing [9, 10]. The analog input bandwidth is a measure of the frequency at which the reconstructed output fundamental drops 3 dB below its low frequency value for a full scale input. The ERBW is defined as the input frequency at which the SNDR drops 3 dB (or ENOB 1/2 bit) below its low frequency value [7]. An ADC used for sub-sampling applications is desired to have an analog input bandwidth and ERBW larger than its Nyquist bandwidth.

2.2.3 Power

Power consumption is also an important parameter of an ADC. It is a primary design constraint for applications that have limited available energy such as devices powered by batteries. Too much power consumption can also lead to a requirement for a heatsink or fan for the IC, which will increase the total system cost. The excessive heat caused by high power dissipation can have negative effect on the reliability of the IC and prevent the integration of more circuit blocks on the same die. Consequently, most designs nowadays are trying to either maximize the performance under a certain power budget or minimize the power consumption for a target performance.

2.2.4 ADC Figure-of-Merit

Various ADC parameters (including parameters mentioned above and others) can be combined to get one single number for the purpose of evaluating ADCs for a certain product or comparing scientific achievement. Numerous ADC figures-of-merits (FOMs) have been proposed and a classification of them can be found in [11]. Two most widely used ADC FOMs in scientific publications are the 'Walden FOM' (FOM_1) and the 'Schreier FOM' (FOM_2) [12, 13]:

$$FOM_1 = \frac{P}{\min\{f_s, 2 \times ERBW\} \times 2^{ENOB}} \tag{2.3}$$

$$FOM_2 = SNDR(dB) + 10log_{10}\left(\frac{BW}{P}\right) \tag{2.4}$$

If FOM_2 is rewritten in linear form and inverted, it is then proportional to

$$\frac{P}{BW \times 2^{2 \times ENOB}} \tag{2.5}$$

which becomes the so called "Thermal FOM" [14]. Comparing Eqs. 2.3 and 2.4, we can clearly see the difference lies in the relative weight given to the conversion accuracy performance. Equation 2.3 implies that the power consumption increases by 2 times when doubling the conversion accuracy (one extra ENOB) which is based on curve-fitting of empirical data [12]; while Eqs. 2.4 and 2.5 account for the fact that due to thermal noise limitations, achieving twice the conversion accuracy requires 4 times increase of the power consumption.

2.3 ADC Performance Limitations and Trends

As illustrated in Fig. 2.5, key factors that influence the ADC performance (in terms of bandwidth, accuracy, and power consumption) are the process technology, ADC architecture, circuit design techniques, and signal/system properties. Limitations of the available process technology, such as minimum feature size, reliability issues, intrinsic capacitance, as well as device imperfections (leakage, mismatch, noise, nonlinearity, etc.), require proper ADC architectures and innovative circuit design techniques to reduce their impacts on the ADC performance.

There is also a trade-off between conversion accuracy, bandwidth and power in designing ADCs using any process technology, improving one of the ADC parameters will mostly likely result in degradation of the other two parameters [15, 16]. The challenge lies in improving all these parameters simultaneously. As discussed in Sect. 2.2, the conversion accuracy of an ADC is limited by many error sources. For those static errors and some of the dynamic errors, numerous

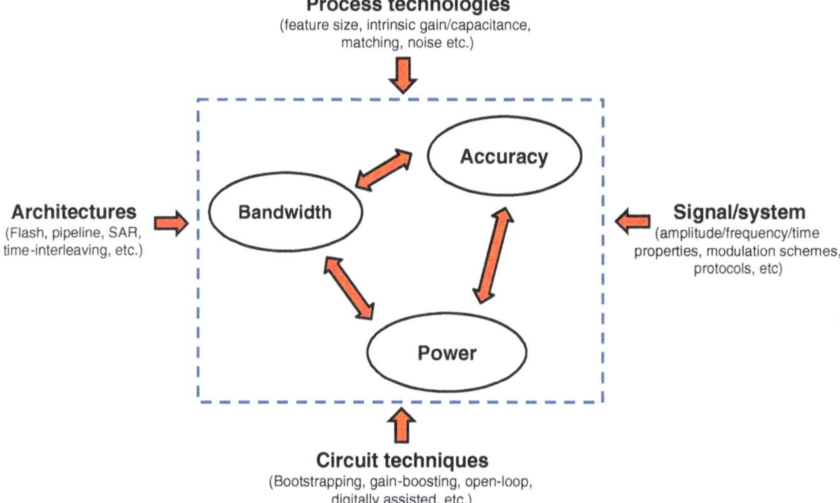

Fig. 2.5 Trade-off between conversion accuracy, speed and power in ADC design and key factors' impact on the overall ADC performance

calibration techniques have been developed to minimize them with little degradation of other ADC performance parameters. Many calibration techniques nowadays exploit the digital signal processing capabilities to "assist" analog circuits of the ADC for accuracy and bandwidth improvement with lower overall power consumption [17, 18]. These techniques measure and correct imperfections of devices and circuits, and they are able to improve the conversion accuracy or sampling speed of an ADC with smaller power overhead compared to the ones without using these techniques. ADC calibrations can be done at startup or in the background without affecting normal operation. However, when the conversion accuracy is limited by random noise, such as thermal noise and clock jitter, improving the conversion accuracy relies on using larger devices to minimize the noise power or increasing the converted signal power. The approach of using larger devices, which refers to the conventional approach mentioned in this book, increases the capacitive loading of the circuit nodes and leads to a higher power consumption for achieving a targeted bandwidth. The required power would actually quadruple per bit increase to maintain the same bandwidth by using this approach to lower the thermal noise power [19]. When the conversion accuracy is limited by quantization noise, the oversample and average technique can be used to improve the conversion accuracy effectively [20], but it requires the ADC to operate at a sampling rate significantly higher than the bandwidth of the signal. When the sampling speed of an ADC exceeds a certain limit of operation frequency, linear increase of the sampling rate further requires an exponential increase of its power consumption [21]. Therefore, the conversion accuracy and bandwidth limitations of an ADC are mainly set by thermal noise, clock jitter and intrinsic capacitance of devices.

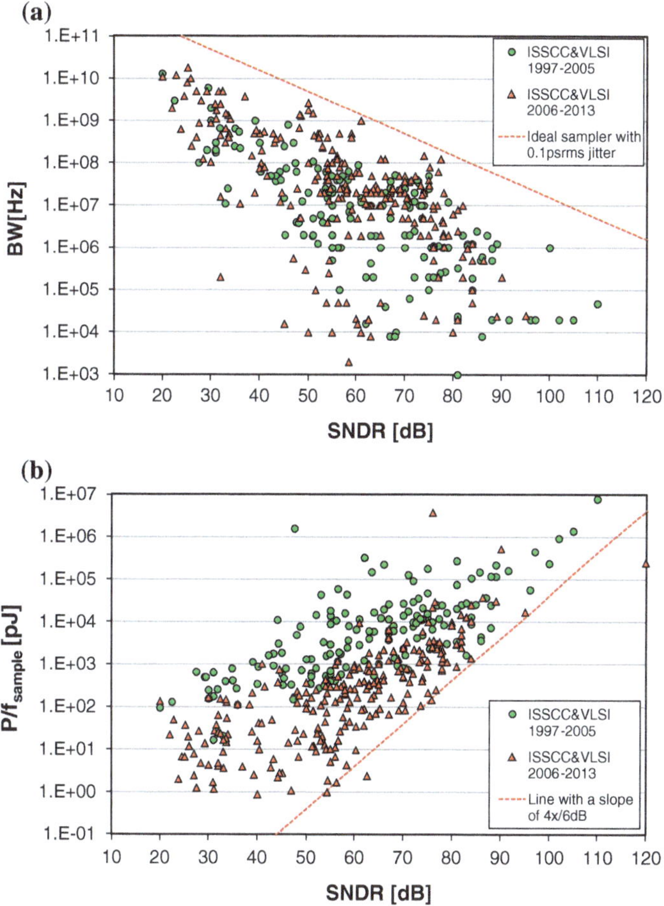

Fig. 2.6 **a** ADC BW versus SNDR, **b** ADC energy efficiency (P/f_{sample}) versus SNDR. The experimental data includes ADCs published in ISSCC and VLSI symposium between 1997 and 2013 [22]

In the following section, recent published state-of-the-art ADCs are studied to find the current performance boundary set by available process technologies, circuit design techniques and architectures. The experimental data used for this purpose includes ADCs published in ISSCC and VLSI Symposium between 1997 and 2013 [22].

Figure 2.6a plots the bandwidth of the ADCs against the SNDR. From this figure, we can see that the achievable bandwidth of the ADCs decreases with higher SNDR. We can also observe that there exists a practical boundary for the achievable bandwidth of state-of-the-art ADCs at different SNDR. As shown in Fig. 2.6a, this boundary is close to the dashed line that represents the performance of an ideal sampler with 0.1 ps clock jitter. ADCs data points close to this line represent what is at present feasible to design. It also implies that having a data point above this line is very challenging or simply not yet feasible with current technologies and techniques. This confirms one of the challenges mentioned above of achieving both higher conversion accuracy and higher bandwidth at the same time.

Figure 2.6b plots the energy efficiency against the SNDR. From this figure, we can observe that there also exists a practical boundary of the achievable conversion efficiency at different SNDR. For ADCs with SNDR greater than 55 dB, this boundary follows a dashed line with a slope of close to 4 times per 6 dB which is the so called 'architecture frontier' [23]. The slope of the dashed line corresponds to the fundamental thermal energy trade-off (power quadruples per 6 dB increase in SNR), and ADCs near this line shown in the figure tend to be thermal noise limited. Observed from this figure, the energy efficiency of the state-of-the-art ADCs with SNDR less than 55 dB on the boundary stay almost the same ($\sim$ 1pJ). With advance in technology, circuit design and architecture innovation, future ADCs with low SNDR will also become thermal noise limited design and get close to the dashed line.

Comparing ADCs published before and after 2006, we observe a slow improvement in the bandwidth-conversion-accuracy product in Fig. 2.6a and a substantial improvement in the energy efficiency of ADC in Fig. 2.6b. The energy efficiency has improved by about 100 times over the last 8 years for ADCs with low SNDR (less than 60 dB). This is mostly enabled by the continuous down scaling of the process technology (minimize device and wiring intrinsic capacitance) and innovations in circuit techniques. As current state-of-art ADCs with SNDR higher than 55 dB are mostly limited by thermal noise, the energy efficiency of these ADCs does not benefit from the process technology scaling due to the lower supply voltage [20, 23].

As observed from publications, state-of-the-art ADCs are well optimized nowadays. To meet the ever-increasing demand for better conversion accuracy, bandwidth and power efficiency, further improvements need to be achieved from process technology improvements, new circuit design techniques, innovative architectures, or signal/system-aware design approaches. Low-to-moderate resolution and high-speed ADCs will continuously benefit from the down-scaling and better optimized process technology (e.g. SOI, FinFET) until they are also limited by thermal noise. For thermal noise limited ADCs, innovative architectures and circuit design techniques to boost the input signal range are an effective way to improve both the bandwidth and energy efficiency which will be discussed in detail in Chap. 3.

In the following sections, an overview of classical ADC architectures is given and an ADC design approach based on exploiting the signal and system properties is also discussed.

2.4 ADC Architectures

Many ADC architectures have been developed over the years. In general, ADCs are divided into two broad categories: Nyquist-rate ADCs and over-sampling ADCs (mainly referred to sigma-delta modulator ADCs). The Nyquist-rate ADCs are the

Table 2.1 Classification of Nyquist-rate ADC architectures

Algorithms	ADC architectures
Parallel search	Flash ADC
Sequential search	Folding ADC
• Linear search	Integrating ADC (single/multi-slop)
• Binary search	Successive approximation ADC
• Sub-binary search	Cyclic ADC
	Sub-ranging ADC
	Pipeline ADC

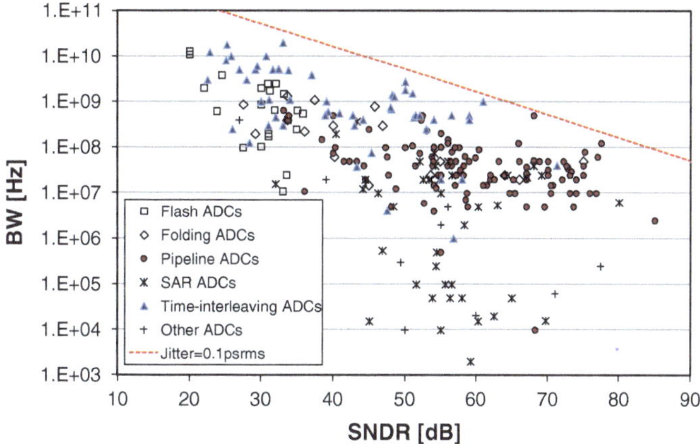

Fig. 2.7 Performance space of different Nyquist-rate ADC architectures (The experimental data includes ADCs published in ISSCC and VLSI symposium between 1997 and 2013 [22])

main focus of this book. Popular Nyquist-rate ADC architectures are listed in Table 2.1 and performance space of different Nyquist-rate ADC architectures is shown in Fig. 2.7. They can be categorized into two groups from the algorithmic point of view, namely parallel search ADCs and sequential search ADCs, or a combination of the two.

The main advantage of parallel search ADCs is in conversion speed, while for sequential search ADCs, their main advantage lies in hardware efficiency, which leads to smaller area for a similar conversion accuracy target [23]. Multiple parallel or sequential search ADCs can be placed in parallel and operate in a time-interleaving fashion to increase sampling speed which refer to the time-interleaving ADC architecture [24]. In the physical implementation, the vast variety of ADC architectures is realized by some basic circuit building blocks, such as track-and-hold, amplifiers, comparators, and reference circuit.

There are some important factors to be considered when comparing different ADC architectures for applications with certain performance requirements, which are conversion time (latency), design complexity, area and power [23].

Conversion time is defined as the time an ADC takes to complete a conversion; it is also specified as latency. The conversion time of a flash ADC does not change with the increase of the number of voltage levels it needs to distinguish. The latency of a SAR ADC or pipelined ADC (1-bit per stage) increases linearly with the increase of its number of bits of resolution. For an integrating ADC, the conversion time increases exponentially with the increase in its number of bits of resolution.

The design complexity of an ADC with certain architecture varies with its performance requirements. Figure 2.7 shows the performance space in terms of bandwidth and SNDR of different ADC architectures based on empirical data. In general, Flash ADCs are suitable architecture for high bandwidth and low resolution applications, SAR ADCs for low bandwidth and moderate-to-high resolution applications, Pipeline ADCs for moderate bandwidth and moderate-to-high resolution applications, and Time-interleaving ADCs for low-to-moderate resolution and very high bandwidth applications. From Fig. 2.7, we can also observe that the performance space of different ADC architectures have a good deal of overlap, this means that multiple architectures can be suitable to meet the target requirements. It is also possible to extend the performance space of certain architectures, but the complexity to design their circuit building blocks to meet the target performance would increase substantially and one architecture may become less competitive compared to other architectures. For example, the flash ADC architecture is suitable for applications requiring very high sampling speeds and low latency but with low resolution. However, selecting the flash ADC architecture to build an ADC with 12 bits resolution and moderate bandwidth is not appropriate. As the number of comparators, the requirements on the comparators and reference, and the associated input capacitance increases exponentially with every additional bit, the difficulty to maintain a large bandwidth and reduce the effect of stronger kick back will result in high design complexity. Instead, achieving such performance with the pipeline architecture is less challenging. Calibration techniques can be used effectively to extend the performance space of certain architectures, but the complexity of calibration circuits increases with the higher performance requirements which should be carefully considered.

Power consumption and die size are also important factors of choosing ADC architectures. For flash converters, every bit increase in resolution requires about 8 times increase in the die size of the ADC core circuitry (number of comparators doubles and each comparator quadruples in size to meet matching requirement). Consequently, the power of the ADC will also increase by 8 times. In contrast, the die size of a SAR, pipelined, or sigma-delta ADC increases linearly with an increase in resolution; while for an integrating ADC, its core die size will not change with an increase in resolution. It is well known that the increase in die size and power consumption increases cost. Trimming and calibration can be used to improve die size and energy efficiency as explained in the previous section. The minimum power required to achieve a certain conversion accuracy and sampling frequency will eventually be limited by thermal noise and clock jitter.

2.5 Exploiting Signal Properties

As discussed in Sect. 2.2, designing ADCs with high conversion accuracy, high sampling speed and low power consumption at the same time is challenging. Recent publications show slow improvement of the ADC performance as today's state-of-the-art ADCs are highly optimized, this is due to limitations of current available process technologies, circuit design approaches and architectures. In order to cope with the ever increasing demand for better ADC performance, it is worthwhile to exploit alternative design approaches. One promising approach is the so called 'signal-aware', 'system-aware' or 'application-aware' design approach [25–27]. Since most of the ADCs nowadays are designed for a specified application, there is much a priori knowledge of the signal and the system available. For example, in communication systems, how source data is encoded and modulated are normally known in advance. This knowledge can be exploited for the design of an optimized ADC for a target application. There are two advantages of this approach:

- The power consumption of the ADC can be reduced without compromising system performance by tailoring the ADC performance to the signal/system properties;
- A better system performance can be enabled without the need of a better ADC which may not be available currently.

Main purpose of the ADC is to digitize the information-bearing waveforms with minimum loss of the information it is intended to convey. The information that needs to be extracted is embedded in one or more properties of the analog waveform such as amplitude, frequency, and phase; the waveform may be corrupted by noise and interfering signals during transmission. Therefore, the a priori knowledge of some properties of the signal waveform (e.g. their probability density function, sparsity, time activities) can be exploited and mapped to the performance requirements of the ADC where opportunities can be found.

The idea of exploiting signal properties to optimize the design of ADCs has been applied to various previous works and shows promising results. In the following sections, various ADC architectures that utilized signal information to improve performance are introduced. Analysis and summary of these existing solutions are given.

Amplitude properties

ADCs are normally designed with uniformly distributed quantization levels. This is only optimal (in terms of quantization noise) when the input signal amplitudes are uniformly distributed. For many applications, the signal amplitude distribution is far from uniform. When knowledge of the amplitude probability distribution function of the signal is available, the quantization levels in the ADC can be optimized according to the probability distribution function of the signal amplitudes to reduce quantization noise [28, 29]. The resulting ADC will have non-uniform distributed quantization levels, having finer quantization for signal amplitudes that have higher probability of occurrence to improve the overall signal-to-quantization-noise-ratio

(SQNR) and bit-error-rate (BER) with a given number of quantization levels. The Lloyd-Max's algorithm was presented in [28] to find the optimal set of quantization thresholds to minimize quantization noise. This approach can be very useful for low resolution ADCs where the quantization noise is the dominant noise source.

Another example of exploiting the signal amplitude property is the 'companding ADC' [30]. As shown in Fig. 2.8, it is realized with three functional building blocks: a signal compressor, a conventional ADC, and a signal expander that inverts the compressor function. With this architecture, a conventional ADC can be used instead of designing an ADC with non-uniform distributed quantization levels to achieve the same function. Ideally, the signal amplitude distribution can be converted into a uniform distribution by the 'compressor' to exploit the dynamic range of the ADC optimally, and after the conversion by the ADC, the signal is restored in the digital domain by the 'expander'. In this way, the restored signal can have a higher SQNR as well as dynamic range compared to a conventional ADC with the same amount of quantization levels for an input signal with non-uniformly distributed amplitudes. In practice, designing a 'compressor' which has a stable non-linear transfer function and achieving good matching between the analog 'compressor' and digital 'expander' is very challenging, therefore a piecewise linear approach is normally adopted [31].

Spectral properties

In many applications, the signals of interest can have large sparsity in the frequency domain which means the actual spectrum occupied by signals is much smaller than the total bandwidth of the spectrum needed to capture at any given time instant. In these situations, sampling at two times the highest signal frequency is inefficient. Such signals can be reconstructed (via a compressed sensing algorithm) with significantly fewer samples than with Nyquist sampling [32]. Therefore, the average sampling rate of the ADC can be relaxed and the amount of output data is reduced. This approach has been demonstrated in various works [33, 34]. For example, [34] applied this approach to build a sampler for wideband spectrally-sparse environments and demonstrated the capability of digitizing an 800 MHz to 2 GHz band with an average sample rate of only 236 Msps which greatly reduced the sample rate requirement of the ADC and power consumption.

Another example of exploiting the spectral properties to enhance the ADC performance is an ADC architecture employing interference detection and cancellation. A mixed-signal architecture with a 'forward interference rejection' approach is presented in [35], which is suitable for processing a weak signal with strong interferences as shown in Fig. 2.9. This architecture contains two low dynamic

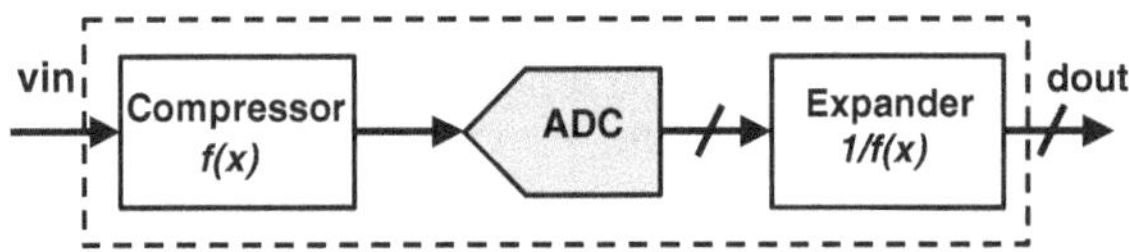

Fig. 2.8 Block diagram of a "companding ADC"

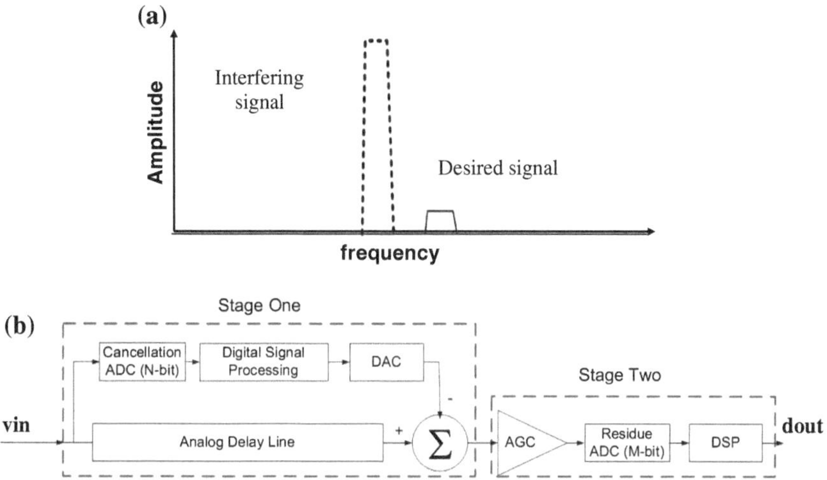

Fig. 2.9 **a** Spectrum of a weak desired signal coexisting with a strong interfering signal, **b** ADC architecture employing interference detection and cancellation [35]

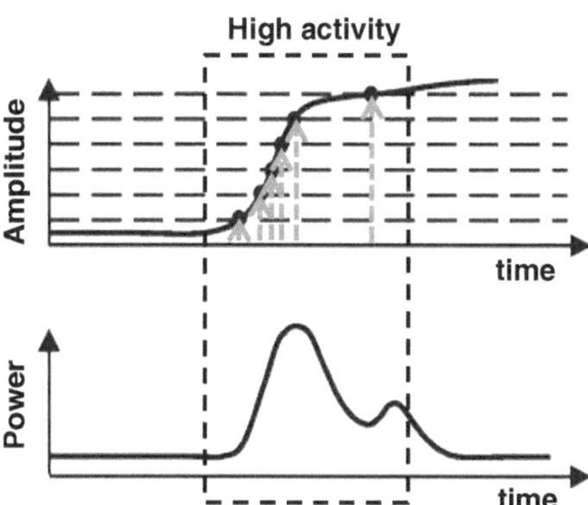

Fig. 2.10 Event-based sampling and estimation of corresponding ADC power dissipation [39]

range ADCs and they can effectively act as a high dynamic range ADC in terms of the ability to resolve a small desired signal in the presence of a large interfering signal. By solving some practical implementation issues of this architecture (delay matching between two signal paths in stage one, signal subtraction and reconstruction, etc.), low dynamic range ADCs can be used to achieve the required system performance, which would otherwise require a high dynamic range ADC and consume significant more power. A programmable notch filter (with control circuitry) can also be used in stage one to achieve the same purpose [36].

Time domain properties

ADCs are normally designed to sample at a constant rate which is based on the worst possible case of the considered applications. Rather than sampling the signals at a constant high rate, the ADC can be designed to adapt its sampling rate according to the activity of the signal [37–41]. Therefore, the power consumption of the ADC can become proportional to the activity of the analog input, as illustrated in Fig. 2.10. For input signals that have burst-like properties in time domain, such as ECG signals, ultrasound signal and UWB impulse signals, significant power can be saved. This approach has been demonstrated in [39–41]. This type of ADCs is commonly referred to as a ‘level-crossing’ or ‘event-driven’ ADC [38, 41].

2.6 Conclusion

In this chapter, the analog-to-digital conversion process and ADC parameters were discussed. The ADC performance limitations, trade-offs between key ADC parameters (conversion accuracy, bandwidth and power consumption), and ADC performance trends were addressed.

As today’s state-of-the-art ADCs are highly optimized due to current available process technologies, circuit design approaches and architectures, it is a challenge to keep pace with the ever increasing demand for more advanced ADCs. However, as most of the ADCs nowadays are designed for a specific application, it is worthwhile to exploit signal and system properties which are a priori knowledge to further enhance the ADC performance. We conclude that this so-called ‘signal-aware’, ‘system-aware’ or ‘application-aware’ ADC design approach, as discussed in Sect. 2.5, is promising for this purpose.

In the following chapters of this book, this concept will be applied to the design of power efficient ADCs for broadband multicarrier systems. In Chap. 3, statistical amplitude properties of multi-carrier signals are exploited and a parallel-sampling ADC architecture for broadband multi-carrier signals is introduced and analyzed.

References

1. van de Plassche, R.J. 2003. *CMOS integrated analog-to-digital and digital-to-analog converters*, 2nd ed. Boston: Springer.
2. Kester, W.A. 2005. *Data conversion handbook*. Amsterdam: Elsevier.
3. Das, A. 2010. *Digital communication: principles and system modeling*. Berlin: Springer.
4. Abidi, A. 2007. The path to the software-defined radio receiver. *IEEE Journal of Solid-State Circuits* 42(5): 954–966.
5. van Roermund, A.H.M. 2007. Smart, flexible, and future-proof data converters. In *18th European conference on circuit theory and design*. pp. 308–319.
6. Gustavsson, M., J.J. Wikner, and N. Tan. 2000. *CMOS data converters for communications*, 2000th ed. Boston: Springer.
7. IEEE standard for terminology and test methods for analog-to-digital converters. *IEEE Std 1241-2010 (Revision of IEEE Std 1241-2000)*, pp. 1–139, Jan 2011.

8. A. N. S. Institute, S. of C. T. Engineers, and G. E. D. (Firm). 2011. *American national standard: measurement procedure for noise power ratio.* Society of Cable Telecommunications Engineers.
9. Nyquist, H. 2002. Certain topics in telegraph transmission theory. *Proceedings of the IEEE* 90 (2): 280–305.
10. Shannon, C.E. 1998. Communication in the presence of noise. *Proceedings of the IEEE* 86(2): 447–457.
11. Jonsson, B.E. 2011. Using figures-of-merit to evaluate measured A/D-converter performance. In *Proceedings of 2011 IMEKO IWADC and IEEE ADC forum*, Orvieto, Italy.
12. Walden, R.H. 1999. Analog-to-digital converter survey and analysis. *IEEE Journal on Selected Areas in Communications* 17(4): 539–550.
13. Schreier, R. and G.C. Temes. 2004. *Understanding delta-sigma data converters*, 1st ed. N. J. Piscataway, and N.J. Hoboken. Chichester: Wiley-IEEE Press.
14. Bult, K. 2006. The effect of technology scaling on power dissipation in analog circuits. In *Analog circuit design*, ed. M. Steyaert, J.H. Huijsing, and A.H.M. van Roermund, 251–294. Netherlands: Springer.
15. Uyttenhove, K., and M.S.J. Steyaert. 2002. Speed-power-accuracy tradeoff in high-speed CMOS ADCs. *IEEE Transactions on Circuits and Systems II: Analog and Digital Signal Processing* 49(4): 280–287.
16. Toumazou, C., Moschytz, G.S., Gilbert, B., and & 0 more. 2004. *Trade-offs in analog circuit design: the designer's companion*ed, 2002 ed. Dordrecht: Springer.
17. Chiu, Y. 1908. Recent advances in digital-domain background calibration techniques for multistep analog-to-digital converters. In *9th international conference on solid-state and integrated-circuit technology, 2008. ICSICT 2008*, pp. 1905–1908.
18. Murmann, B., C. Vogel., and H. Koeppl. 2008. Digitally enhanced analog circuits: system aspects. In *IEEE international symposium on circuits and systems, 2008. ISCAS 2008*, pp. 560–563.
19. Murmann, B. 2010. Trends in low-power, digitally assisted A/D conversion. *IEICE Transactions on Electronics* E93-C(6): 718–729.
20. Temes, G.C., and J.C. Candy. 1990. A tutorial discussion of the oversampling method for A/D and D/A conversion. In *IEEE international symposium on circuits and systems, 1990*, vol. 2, pp. 910–913.
21. Dusan, Stepanovic. 2012. *Calibration techniques for time-interleaved SAR A/D converters.* Ph.D. Dissertation. EECS Department, University of California, Berkeley.
22. Murmann, B. ADC performance survey 1997–2013. http://www.stanford.edu/~murmann/adcsurvey.html.
23. Pelgrom, M.J.M. 2012. *Analog-to-digital conversion*, 2nd ed. 2013 edition. New York: Springer.
24. Black, J.W.C., and D. Hodges. 1980. Time interleaved converter arrays. *IEEE Journal of Solid-State Circuits* 15(6): 1022–1029.
25. Lin, Y., K. Doris, H. Hegt, and A.H.M. Van Roermund. 2012. An 11b pipeline ADC with parallel-sampling technique for converting Multi-carrier signals. *IEEE Transactions on Circuits and Systems I: Regular Papers* 59(5): 906–914.
26. Shanbhag, N., and A. Singer. 2011. System-assisted analog mixed-signal design. In *Design, automation test in Europe conference exhibition (DATE), 2011*, pp. 1–6.
27. Murmann, B. 2013. Digitally assisted data converter design. In *Proceedings of the ESSCIRC (ESSCIRC) 2013*, pp. 24–31.
28. Max, J. 1960. Quantizing for minimum distortion. *IRE Transactions on Information Theory* 6 (1): 7–12.
29. Lloyd, S. 1982. Least squares quantization in PCM. *IEEE Transactions on Information Theory* 28(2): 129–137.
30. Tsividis, Y. 1997. Externally linear, time-invariant systems and their application to companding signal processors. *IEEE Transactions on Circuits and Systems II: Analog and Digital Signal Processing* 44(2): 65–85.

31. Maheshwari, V., Serdijn, W.A., and J.R. Long. 2007. Companding baseband switched capacitor filters and ADCs for WLAN applications. In *IEEE international symposium on circuits and systems, 2007. ISCAS 2007*, pp. 749–752.
32. Candes, E.J., and M.B. Wakin. 2008. An introduction to compressive sampling. *IEEE Signal Processing Magazine* 25(2): 21–30.
33. Chen, X., Z. Yu, S. Hoyos, B.M. Sadler, and J. Silva-Martinez. 2011. A sub-nyquist rate sampling receiver exploiting compressive sensing. *IEEE Transactions on Circuits and Systems I: Regular Papers* 58(3): 507–520.
34. Wakin, M., S. Becker, E. Nakamura, M. Grant, E. Sovero, D. Ching, J. Yoo, J. Romberg, A. Emami-Neyestanak, and E. Candes. 2012. A nonuniform sampler for wideband spectrally-sparse environments. *IEEE Journal on Emerging and Selected Topics in Circuits and Systems* 2(3): 516–529.
35. Yang, J. 2010. *Time domain interference cancellation for cognitive radios and future wireless systems*. PhD diss., EECS Department, University of California, Berkeley.
36. Silva-Martinez, J., and A.I. Karşılayan. 2010. High-performance continuous-time filters with on-chip tuning. In *Analog circuit design*, ed. A.H.M. Roermund, H. Casier, and M. Steyaert, 147–166. Netherlands: Springer.
37. Mark, J.W., and T.D. Todd. 1981. A nonuniform sampling approach to data compression. *IEEE Transactions on Communications* 29(1): 24–32.
38. Tsividis, Y. 2010. Event-driven data acquisition and digital signal processing—a tutorial. *IEEE Transactions on Circuits and Systems II: Express Briefs* 57(8): 577–581.
39. Weltin-Wu, C., and Y. Tsividis. 2013. An event-driven clockless level-crossing ADC with signal-dependent adaptive resolution. *IEEE Journal of Solid-State Circuits* 48(9): 2180–2190.
40. Kim, H., C. van Hoof, and R.F. Yazicioglu. 2011. A mixed signal ECG processing platform with an adaptive sampling ADC for portable monitoring applications. In *2011 annual international conference of the IEEE engineering in medicine and biology society*, pp. 2196–2199.
41. Kozmin, K., J. Johansson, and J. Delsing. 2009. Level-crossing ADC performance evaluation toward ultrasound application. *IEEE Transactions on Circuits and Systems I: Regular Papers* 56(8): 1708–1719.

Chapter 3
Parallel-Sampling ADC Architecture for Multi-carrier Signals

Abstract This chapter starts with a brief introduction of broadband multi-carrier transmission in Sect. 3.1. Section 3.2 describes the amplitude properties of multi-carrier signals, especially their large peak-to-average ratio. A discussion of the ADC dynamic range requirement for a multi-carrier system is given in Sect. 3.3. Section 3.4 reviews power reduction techniques to enhance the SNR of noise limited ADCs in advanced CMOS technologies. Section 3.5 presents a parallel-sampling architecture for ADCs to convert multi-carrier signals efficiently by exploiting their amplitude statistical properties. ADCs with this architecture are able to have a larger input signal range without causing excessive distortion while showing an improved accuracy over the small amplitudes that have much higher probability of occurrence due to the multi-carrier signal amplitude properties. The power consumption and area of ADCs with the parallel-sampling architecture can be reduced to achieve a desired SNR for multi-carrier signals compared to conventional ADCs. Section 3.6 proposes four implementation options of the parallel-sampling ADC architecture and Sect. 3.7 concludes the chapter.

3.1 Introduction to Multi-carrier Transmission

The demand for higher data rates in modern digital communication systems is growing today at an explosive pace. Especially, in wireless communication systems such as cellular and WLAN data rates have increased by 100 times over the last decade and another 10 times is projected in the next 5 years, as it was observed in [1]. According to the Shannon–Hartley theorem [2, 3], the data rates can be increased by expansion of the channel bandwidth or improvement of the channel quality. However, the scarcity of licensed spectrum in wireless communication systems and the physical bandwidth limitation of the channel in most of wireline communication systems limit the expansion of bandwidth for data transmission. The other approach to meet this ever increasing demand of data rate is to push the spectral efficiency of the data transmission within the available bandwidth to its limits.

Y. Lin et al., *Power-Efficient High-Speed Parallel-Sampling ADCs for Broadband Multi-carrier Systems*, Analog Circuits and Signal Processing,
DOI 10.1007/978-3-319-17680-2_3

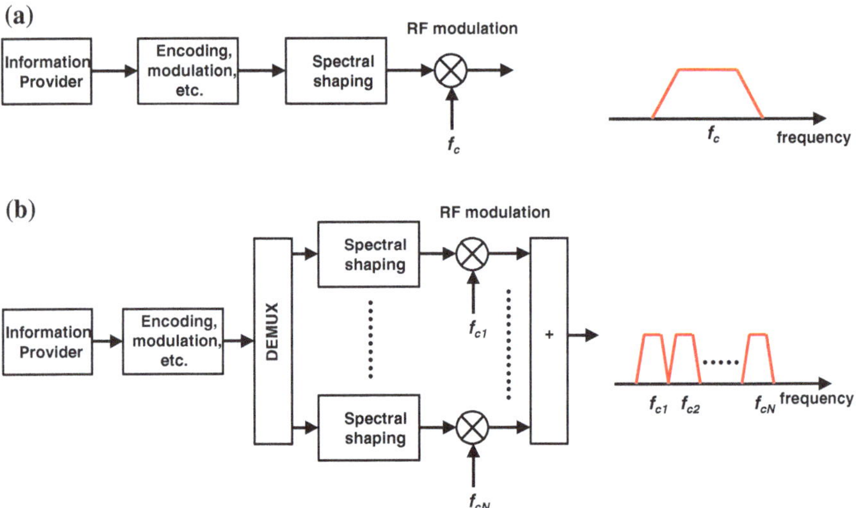

Fig. 3.1 Block diagrams and output spectrum of **a** a typical single-carrier transmitter and **b** a typical multi-carrier transmitter

A widely adopted technique to improve the data rate is multi-carrier transmission. Multi-carrier transmission implies that, instead of transmitting a single broadband signal, multiple narrowband signals are frequency multiplexed and jointly transmitted [4, 5]. Block diagrams of typical single and multi-carrier transmitters and their signal spectrum are shown in Fig. 3.1. There are many advantages of multi-carrier transmission including high spectral efficiency, ability to cope with frequency selective and time dispersive propagation channel conditions without complex equalization filters, and efficient implementation of modulation and demodulation functions in the digital domain by using IFFT and FFT [5]. One special case of multi-carrier transmission is orthogonal frequency division multiplexing transmission (OFDM) which allows sub-carriers to overlap in frequency domain to maximize spectral efficiency [5]. Thanks to the advantages mentioned above, the multi-carrier transmission has been widely adopted in many communication systems, such as digital audio/video broadcasting (DAB/DVB) standards in Europe, high-speed digital subscriber line (DSL) modems over twisted pairs, digital cable television systems, powerline communication, as well as mobile and wireless networks such as 4 G Long Term Evolution (LTE), wireless LAN (IEEE802.11a/g/n), worldwide interoperability for microwave access (WiMAX), and wireless personal area network (WPAN) [5].

A major drawback of multi-carrier transmission is that its transmitted signals exhibit a high peak-to-average power ratio (*PAPR*), which means the signal waveform has instantaneous 'peak' amplitudes much larger than its root mean square (*RMS*) amplitude value. This is an undesired signal property, as it causes performance degradation of the system, power efficiency reduction of the power amplifier, and an increase in circuit/system complexities [6]. In the next section, statistical amplitude properties of multi-carrier signals will be analyzed.

3.2 Statistical Amplitude Properties of Multi-carrier Signals

Figure 3.2 shows a single sinusoid and a multi-carrier signal in time domain. These two signals are plotted with equal *RMS* amplitude values. A multi-carrier signal has different statistical amplitude properties compared to that of a sinusoid signal. Its amplitude exhibits large variations and it has large amplitude 'peaks' when compared to its *RMS* value. This signal property is characterized by the *PAPR* which is defined, in dB, as follows [5]:

$$PAPR_{multi-carrier} = 10\log_{10}\left(\frac{\max|x(t)|^2}{E\{|x(t)|^2\}}\right) \tag{3.1}$$

$$x(t) = \frac{1}{\sqrt{N}}\sum_{n=0}^{N-1} g(t)\cdot d(n)\cdot e^{j\cdot 2\pi\cdot n\cdot f_{cs}\cdot t},\ 0\leq t\leq T_b \tag{3.2}$$

where *x*(*t*) is the superposition of N complex-modulated sinusoidal waveforms, each corresponding to a given subcarrier, E{.} denotes the expectation, *g*(*t*) is the pulse function, *d*(*n*) is the data symbol, *N* is the number of subcarriers, f_{cs} denotes the sub-carrier spacing and $T = 1/f_{cs}$ is the data block duration.

With a large number of modulated sub-carriers, the multi-carrier signal has a large *PAPR* value as can be observed from both Eqs. (3.1) and (3.2). In the extreme situation where all the sub-carriers are correlated, the *PAPR* has a theoretical maximum value equal to $10\log_{10}(N) + PAPR_{sub-carrier}$, where $PAPR_{sub-carrier}$ is the peak-to-average power ratio of the subcarrier. For example, in the IEEE

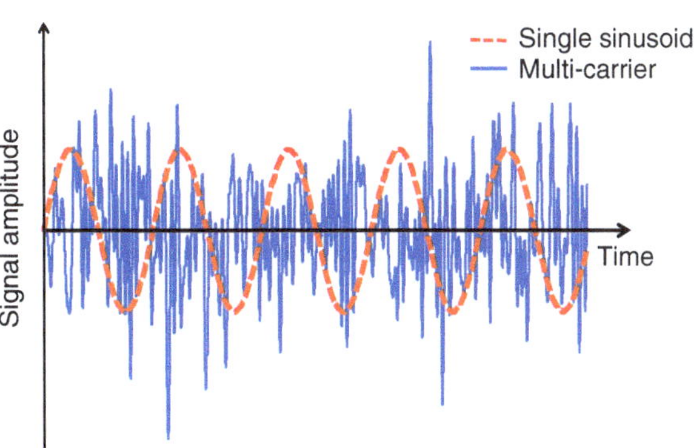

Fig. 3.2 Comparison of a single sinusoid and a multi-carrier signal in time domain

802.11ac system [7], up to 484 256-QAM modulated sub-carriers occupying 160 MHz of spectrum are used to transmit data in parallel. The composed signal has a theoretical maximum *PAPR* of about 34 dB (*PAPR* of a single 256-QAM modulated carrier is about 7.2 dB [9]). In a broadband receiver for a DOCSIS 3.0 cable modem [8], the received signal can be composed by up to 126 256-QAM modulated sub-carriers and has a maximum *PAPR* of more than 28 dB. In reality, the data on different sub-carriers is mostly uncorrelated which means that these sub-carriers have different amplitude and phase values. Therefore the probability of getting the theoretical maximum *PAPR* is very small. When the number of uncorrelated sub-carriers is adequately large, the composed signal waveform has an amplitude distribution approaching the Gaussian distribution according to the Central Limit Theorem [10].

To give a more intuitive view of this signal property, Fig. 3.3 shows the waveforms and histograms of a sinusoid, a single carrier signal with 256-QAM modulation and a multi-carrier signal having 50 256-QAM modulated sub-carriers. The data symbols in both the single carrier and multi-carrier signals are shaped by a square-root raised-cosine filter to minimize intersymbol interference. These three waveforms have the same *RMS* amplitude value of 1 and zero mean value. The histograms of these signals shown in Fig. 3.3 reveal the differences of their amplitude probability distributions. The single sinusoid waveform has a U-shaped probability density function (*pdf*); large amplitudes have higher probability of occurrence compared to small amplitudes and the maximum amplitude level is $\sqrt{2}$ times its *RMS* amplitude value. Both the single 256-QAM modulated signal and the multi-carrier signal have a bell-shaped *pdf*, this implies that small amplitudes have higher probability of occurrence compared to large amplitudes and the probability of occurrence is decreasing with the increase of amplitude value. The *pdf* of the multi-carrier signal has a 'long tail' due to its Gaussian-like amplitude distribution; more than 90 % of the amplitudes are smaller than two times of its *RMS* value, while its peak amplitude level is far larger than its *RMS* value.

In reality, circuit building blocks in any system, such as DAC, power amplifier, analog filter and ADC, have limited input and output signal ranges, as illustrated in Fig. 3.4. Signal amplitudes beyond the maximum range value will be cut off which refers to "clipping". Figure 3.5 plots the probability of clipping with respect to the clipping ratio (*CR*) of three signal waveforms. The *CR* is defined as the ratio between the clipping threshold, which is the maximum range value, and the *RMS* amplitude level. The *PAPR* of the sinusoid and the single 256-QAM modulated signal are about 3 and 9.6 dB respectively. Therefore, clipping can be avoided by choosing a CR higher than these values. As shown in Fig. 3.5, numerical simulation shows that the probability of clipping of a multi-carrier signal follows that of an ideal Gaussian distributed signal. Consequently, a *PAPR* higher than 13 dB can be expected. Even with a *CR* as large as 4.5 ($\sim$13 dB), the clipping probability of a signal with Gaussian amplitude distribution is still higher than 10^{-5}.

The high *PAPR* is an undesired property of the multi-carrier signal. To accommodate the instantaneous large amplitude peaks and avoid excessive signal

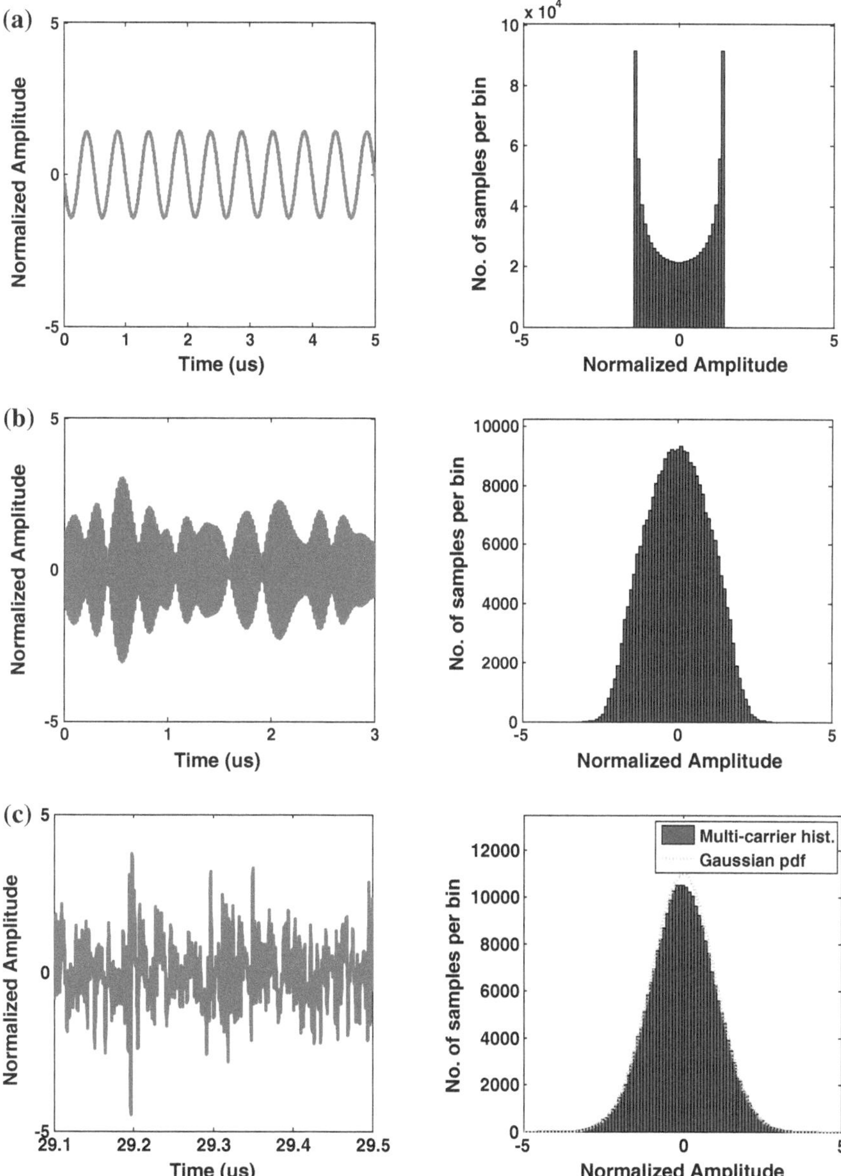

Fig. 3.3 Signal waveforms and histograms: **a** a single sinusoid, **b** a 256-QAM modulated single-carrier and **c** a multi-carrier having 50 256-QAM modulated sub-carriers

distortion due to clipping, the circuit building blocks of the system are required to have a large dynamic range to achieve the desired system performance. The demand of a large dynamic range translates to higher circuit design complexity and higher power consumption.

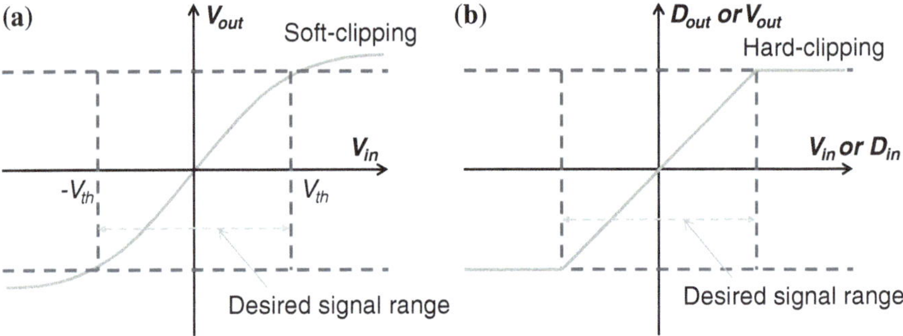

Fig. 3.4 Typical transfer functions of **a** an amplifier and **b** an ADC or DAC

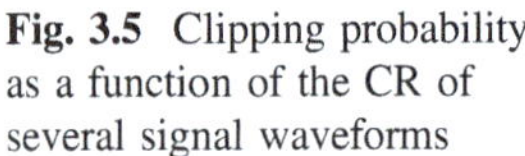
Fig. 3.5 Clipping probability as a function of the CR of several signal waveforms

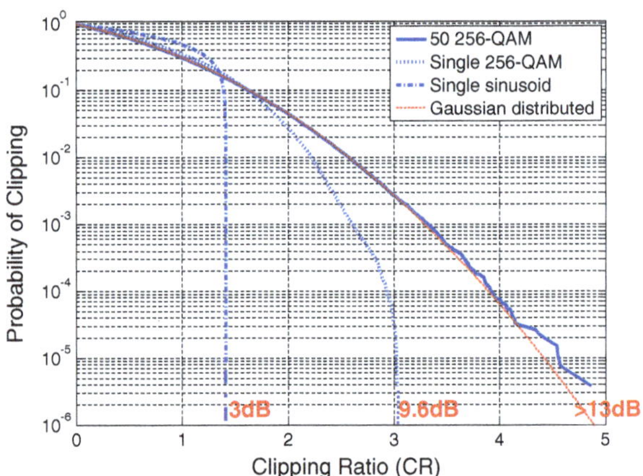

3.3 ADC Requirements for Multi-carrier Signals

Figure 3.6a shows a typical front-end block diagram of a multi-carrier receiver. In such a receiver, the ADC is normally preceded by an RF/analog front-end, which is designed to filter unwanted signals and deliver a well-defined power level to the input of the ADC, and is followed by a DSP unit. The main performance requirements of the ADC are sampling rate, conversion accuracy, and power consumption as discussed in Chap. 2. To meet the ever increasing user demands for higher data rates, many broadband multi-carrier systems nowadays require ADCs with high bandwidth and high resolution and at the same time low power consumption. This make the ADC one of the most challenging circuit building blocks in the system.

An example of a multi-carrier signal before and after it is converted by the ADC is shown in Fig. 3.6b. Two dashed lines denote the input range of the ADC. If the input signal levels exceed the ADC's input range either at the positive or negative

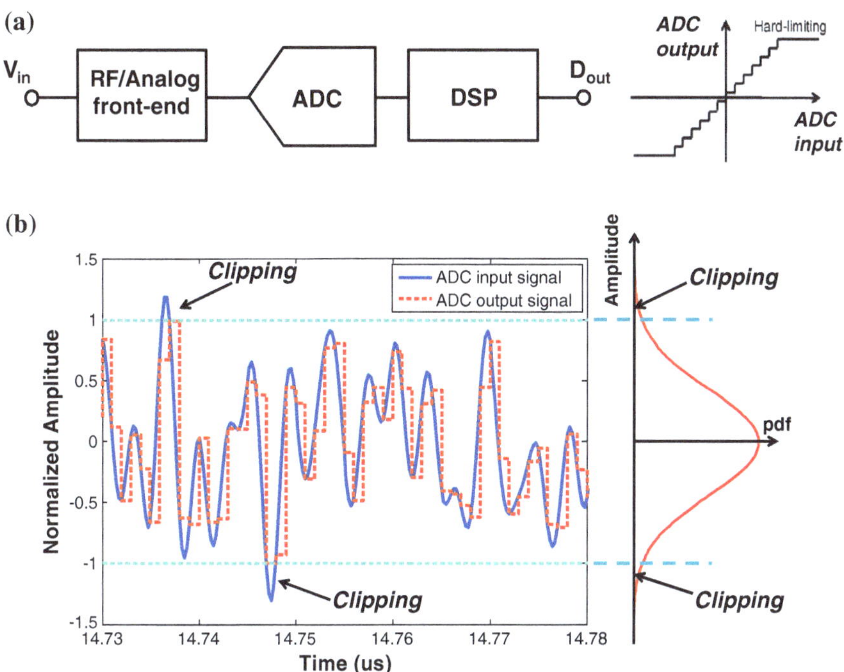

Fig. 3.6 **a** A typical front-end block diagram of a multi-carrier receiver and the ADC transfer curve; **b** a multi-carrier signal before and after it is converted by an ADC

side, the ADC acts as a hard limiter for its input signal and the corresponding ADC's output codes are saturated to its maximum or minimum value, which results in clipping. Clipping is a nonlinear process and causes significant noise increase [11]. It reduces the SNDR of the ADC's output signal and leads to degradation of the performance of the system in terms of BER [11]. Figure 3.7 shows an example of the conversion errors due to quantization and clipping of an 11 b ADC for a broadband multi-carrier signal and the effect of these errors in the frequency domain. The multi-carrier signal used in the example is the one shown in Fig. 3.2c. In this example, the average signal power is 8 dB lower than the ADC's full scale signal power (P_{fs}) which is measured with a full scale sinusoid. However, the conversion errors due to clipping can still be as large as tens of LSBs as shown in Fig. 3.7a, and thus much larger than the quantization errors which in the ideal case are less than ½ LSB. The clipping errors lead to a significant increase of the noise floor. Observed from the output signal spectrum shown in Fig. 3.7b, the total noise power is in this example increased by 16 dB compared to that only caused by quantization.

The conventional way to avoid excessive clipping of the signal is by backing off the signal power by a large factor with respect to P_{fs} [9]. When no clipping is allowed, the required amount of power back-off needs to be at least as large as the

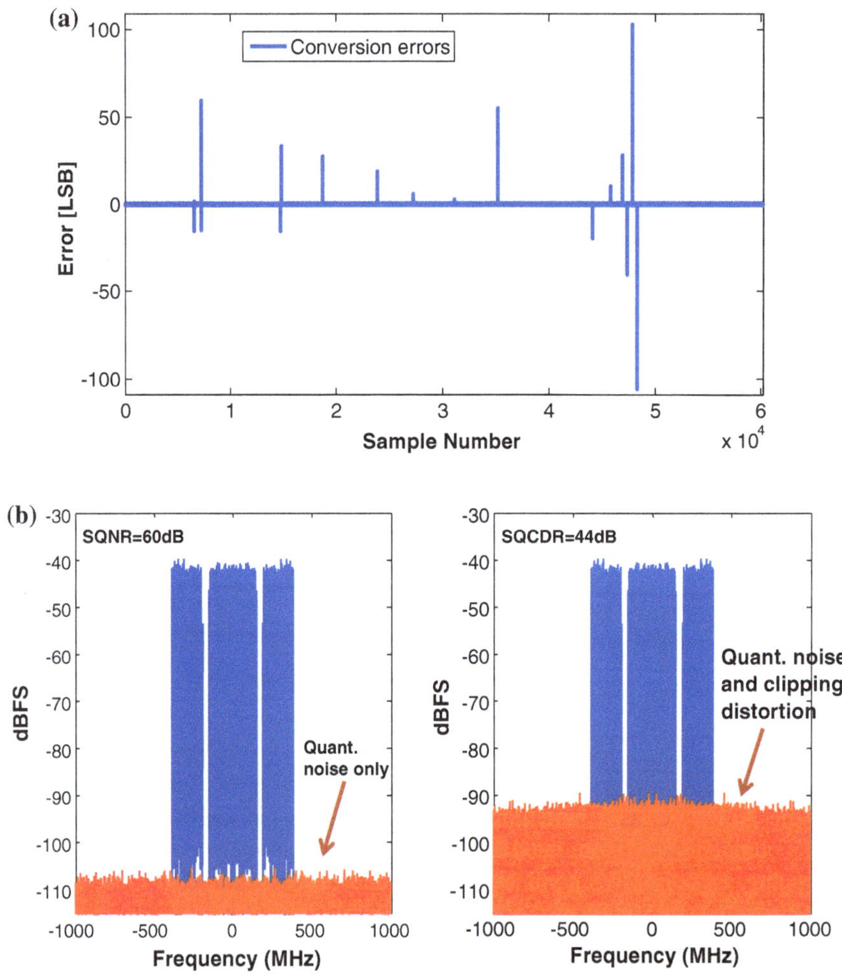

Fig. 3.7 **a** ADC conversion errors due to quantization and clipping ($E(n \cdot T_s)$, n represents the sample number); **b** examples of ADC output spectrum of a multi-carrier signal with 50 256-QAM modulated sub-carriers with quantization noise and with both quantization noise and clipping distortion

PAPR of the signal (the power back-off is defined by P_{fs} over the signal power). For an ADC with a given input signal range and conversion accuracy, the input signal power back-off improves the signal-to-clipping-distortion ratio (SCDR) but degrades the signal-to-quantization-noise ratio (SQNR) as well as signal-to-thermal-noise ratio (STNR) of the ADC output, as the noise power due to quantization noise and thermal noise is independent of the input signal power. This is illustrated in Fig. 3.8.

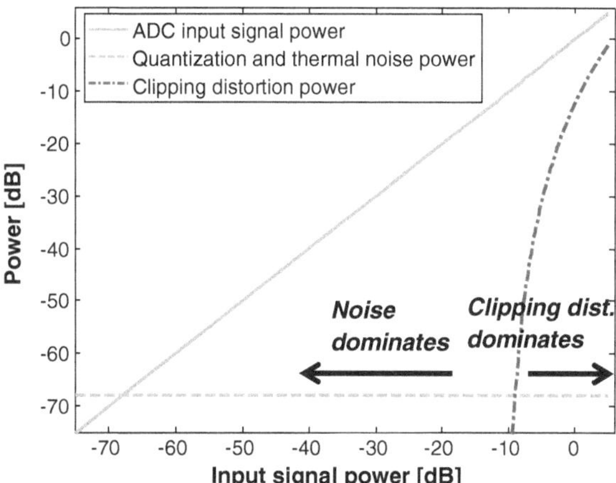

Fig. 3.8 Relation of signal, noise and clipping distortion power of an 11 b ADC converting a signal waveform with Gaussian amplitude distribution

As discussed in the Sect. 3.2, the multi-carrier signal has a Gaussian like amplitude probability distribution and a large *PAPR*. The approach of specifying the dynamic range of the ADC to cover the largest instantaneous amplitude peaks which occur with very low probability is inefficient. Instead, the amplitude statistic property of the signal can be exploited to reduce the dynamic range requirement of the ADC by allowing certain amounts of clipping events while still meeting the desired system requirement. There exists a best tradeoff between clipping distortion and quantization noise that results in the highest signal-to-noise-and-clipping-distortion ratio (SNCDR) [9]. For signals with known amplitude probability distribution, ADC noise power and input range, the optimal input signal power that results in the maximum value of the SNCDR can be found from Eq. (3.3) and the corresponding optimal power back-off value is given by Eq. (3.4).

$$SNCDR = \frac{P_s}{P_n + P_c} = \frac{v_{s.RMS}^2}{v_{n.RMS}^2 + \int_{L_{clip}}^{\infty} (x - L_{clip})^2 f(x)dx + \int_{-\infty}^{-L_{clip}} (x + L_{clip})^2 f(x)dx} \tag{3.3}$$

$$P_{backoff} = 10 \cdot \log_{10}(P_{fs}/P_s) = 20 \cdot \log_{10}(CR) - 3 \tag{3.4}$$

where L_{clip} is the clipping level of the ADC which is equal to its maximum input level; CR is the clipping ratio $L_{clip}/v_{s.RMS}$; $f(x)$ is the probability density function of the signal amplitude with a variance equal to $(L_{clip}/CR)^2$; x denotes the signal amplitude level; $v_{n.RMS}$ is the *RMS* noise voltage of the ADC; $P_{backoff}$ is the power back-off value expressed in logarithmic scale; P_s, P_n, and P_c are the input signal power, total noise power, and clipping distortion power respectively.

To give a more intuitive understanding of the optimal power back-off for signals with different amplitude distribution, Fig. 3.9 plots SNCDR of an 11 b ADC with respect to its input signal power of the three signal waveforms (a single sinusoid,

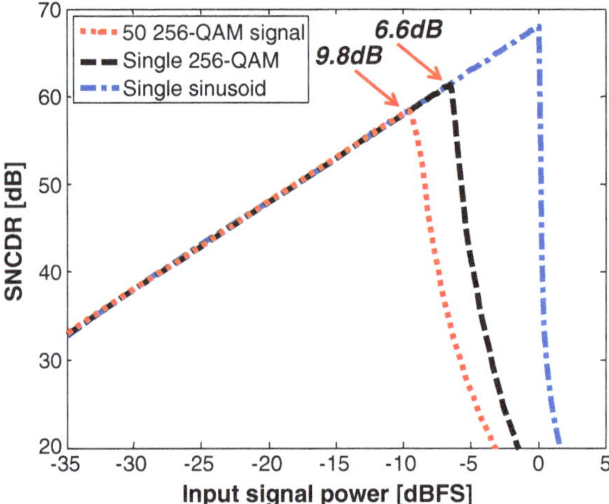

Fig. 3.9 The SNCDR of an 11 b ADC output versus the input signal power of a single sinusoid, a 256-QAM modulated single-carrier signal and a multi-carrier signal having 50 256-QAM modulated sub-carriers. (The input signal power is expressed relative to the power of a full scale sinusoid.)

a 256-QAM modulated single-carrier signal and a multi-carrier signal having 50 256-QAM modulated sub-carriers). For the single sinusoid, the SNCDR degrades significantly as soon as the ADC clips the signal due to its U-shaped amplitude probability distribution, while for the other two signals, due to their amplitude *pdf*, a certain amount of clipping events can be tolerated to have a best compromise between noise and clipping distortion. Instead of more than 13 dB back-off to cover the largest amplitude level of the multi-carrier waveform, a power back-off value of 9.8 dB is found to achieve the maximum SNCDR. This corresponds to a reduction of the ADC dynamic range requirement of more than 3 dB.

From Eq. (3.3), the maximum SNCDR of ADC with different resolutions for a certain signal can be derived. For a signal having a Gaussian amplitude distribution with $f(x) = 1/\sigma\sqrt{2\pi} \cdot e^{-\frac{1}{2}\left(\frac{x}{\sigma}\right)^2}$, $\sigma = L_{clip}/CR$, the SNCDR of ideal ADCs with different resolutions can be found as shown in Fig. 3.10. The maximum SNCDR and the optimal power back-off with the ADC resolution can be found from analytical simulations and their relationship can be found by curve fitting as:

$$SNCDR_{\max} \approx 5.5 \cdot N - 2.62 \tag{3.5}$$

$$P_{optimal\_backoff} \approx -0.04 \cdot N^2 + 1.45 \cdot N - 0.58 \tag{3.6}$$

where N represents the number of bits of an ADC.

From these equations, each extra bit corresponds to about 5.5 dB increase in SNCDR for a signal with Gaussian amplitude distribution. The optimum power back-off values vary with different ADC resolutions which depend on the best

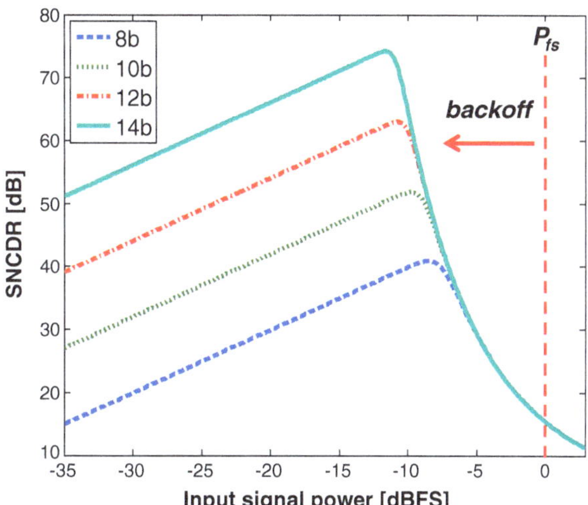

Fig. 3.10 The SNCDR of ideal ADCs with different resolution for a signal with Gaussian amplitude *pdf* versus its input signal power

compromise between the quantization noise and clipping noise. For higher resolution ADCs, less clipping events of the multi-carrier signal are allowed due to the requirement of lower total noise power.

The optimal power back-off approach improves the usage of the ADC's dynamic range for converting the multi-carrier signals and relaxes the ADC's dynamic range requirement for getting the desired conversion accuracy. However, the dynamic range of the ADC is still not well exploited. For example, with 12 dB power back off (corresponding to a CR of 4), half of the ADC's input range is only exploited by less than 3 % of the total signal amplitudes. This results in very inefficient usage of the ADC's dynamic range, offering an opportunity to improve it with innovative techniques.

3.4 Power Reduction Techniques for Thermal-Noise Limited ADCs

The conversion accuracy of an ADC depends essentially on the strength of the signal compared to the errors due to devices mismatch, all sources of noise and distortion. If the ADC conversion accuracy is thermal noised limited, it can be enhanced by boosting the input signal power (requiring a larger linear input signal range) without significantly distorting the signal and/or by reducing the total noise power of the ADC. Improving the SNR by reducing the intrinsic thermal noise power requires larger devices and the associated larger capacitances have negative effects on speed and power consumption of the ADC. For ADCs designed in newer CMOS technology nodes, achieving a large linear signal range is getting more challenging due to the reduction of supply voltages. With the supply voltage down scaling, due to reliability concerns in newer CMOS technologies, many authors predicted that the power efficiency for analog circuits gets lower rather than that it improves for maintaining the same SNR [12–15]. Since the power consumption of high-speed and high-resolution ADCs are normally dominated by their analog circuits, designing these ADCs with better power efficiency in advanced CMOS technologies requires circuit and architecture innovations.

As discussed in Chap. 2, high-speed and high-resolution ADCs nowadays are mostly thermal noise limited. For these ADCs, there exists a strong trade-off between the power consumption and SNR, which can be observed from Eq. (3.7). It is derived from a typical switched-capacitor circuit, as shown in Fig. 3.11, which is the basic building block of an ADC [13–15]:

$$P \propto \frac{k \cdot T \cdot SNR \cdot f_s}{\eta_{vol} \cdot \eta_{cur}} \tag{3.7}$$

where P is the ADC power consumption, $k \cdot T$ is the thermal energy, V_{dd} is the supply voltage, η_{vol} is the voltage efficiency factor which equals the *RMS* value of the input signal amplitude over V_{dd}, η_{cur} is the current efficiency which equals the *RMS* value of the current to charge the load over the *RMS* value of the current drawn from the power supply, SNR is proportional to $(\eta_{vol} \cdot V_{dd})^2/(k \cdot T/C)$, and f_s is the switching frequency.

Considering constant η_{vol}, η_{cur}, and T, it is clear from Eq. (3.7) that increasing the SNR by 6 dB requires a 4 times increase in power dissipation for a given sampling rate. It also indicates that the power can be reduced for a desired *SNR* and f_s by improving η_{vol} and η_{cur}. Many previous works have demonstrated circuit design techniques to reduce the power consumption of ADCs based on improving the η_{cur} of the buffers and amplifiers. These techniques, as summarized in [17, 18], include open-loop amplifiers with digital calibration, amplifier sharing, higher current efficiency amplifier architectures such as class-AB amplifiers, comparator based amplifiers, dynamic amplifiers, and so on.

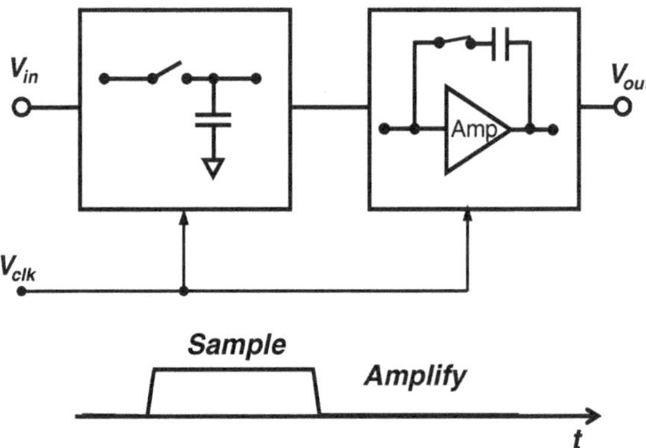

Fig. 3.11 General view of switched-capacitor amplifier [16]

Another effective way to minimize the power consumption is by improving the voltage efficiency (η_{vol}) of ADCs. Enabling the ADC to process a large input signal range linearly allows reducing the sampling capacitor size that determines the thermal noise. The reduction of the capacitors brings many benefits such as higher bandwidth, smaller area, and lower driving current for a given speed. Due to the down scaling of the supply voltage in advanced CMOS technology (e.g. nominal core supply voltage is 0.9 V or lower for 28 nm CMOS technology) [19, 20], achieving a large input signal range is especially important for maintaining good power efficiency for noise limited ADCs.

However, increasing the signal range of an ADC is normally constrained by the linearity of the input sampling stage, the amplifier's output stage, and further by the reference voltage. Several new techniques have been proposed to enlarge the signal range that can be processed by the ADC and have demonstrated their advantages in improving the power efficiency of ADCs comparing to the approach of using larger devices to lower thermal noise. Bootstrapped circuit techniques have been widely used in the sampling stage of low voltage ADCs to enlarge their linear input signal range [21]. Within the ADC, such as low voltage pipeline ADCs, the linear signal range that can be processed is normally limited by the amplifier output swing. In [22, 23], a technique called 'range scaling' or 'dynamic-range-doubling' is used in the first pipeline stage to decouple the choice of the amplifier's input and output signal swing, such that the input and output signal range of the stage can be optimized separately. This technique enabled high-resolution high-speed ADCs with very good power efficiency in advanced CMOS technology with a single, low supply voltage. Another circuit technique that has proven advantage in improving power efficiency of high-speed and high-resolution ADCs is the mixed-supply-voltage design approach [24]. This approach exploits techniques allowing the hybrid use of thick and thin oxide devices to boost circuit performance in advanced CMOS technologies. In [25], thick oxide devices with a high supply voltage are used for the first sampling stage of a pipeline ADC and low supply voltage of the rest of the ADC for lower power consumption. However, using thick oxide devices in the sampling switches introduces higher on-resistance of the switches and large capacitive loading for the clock buffers and results in degradation of the achievable sampling speed. In [26], a 5.4 GS/s, 12 b pipeline ADC with high power efficiency was implemented in 28 nm CMOS technology by exploiting the benefit of using a combination of thick and thin oxide devices working at multiple supply voltages. The thin oxide devices are used for the sampling stage and comparators that require fast operation as well as the input stage of the amplifier that requires high g_m with low parasitic capacitance. The amplifiers are supplied by a high supply voltage (2.5 V) instead of the nominal supply voltage (1.1 V) to achieve both high gain and large linear output signal range at the same time. A similar approach was applied in a 3 GS/s 11 b TI-SAR ADC which achieved state-of-the-art performance [27]. In this work, thin oxide devices were used for the switches in the signal path to achieve fast operation, while the buffer and DAC used a combination of thin and thick oxide devices and operated at a high supply voltage to achieve both a large output signal range and high speed. However, the achievable input signal range is still constrained by the input sampling stage which needs to process the whole signal range linearly.

The power reduction techniques for thermal noise limited ADCs discussed above are mostly at circuit and architecture levels. The power reduction can also be achieved through innovations in technology/devices as well as at system level. For example, technology/devices can be engineered to have lower excess noise factor, low intrinsic devices capacitance, or higher break down voltage to accommodate a large signal swing without reliability issues. As discussed in Chap. 2, certain signals and systems properties of an application also offer opportunities to develop power reduction techniques for ADCs, such as event driven sampling and compressive sensing to reduce the average power consumption of an ADC for a desired performance [28, 29].

By exploiting the statistical signal amplitude properties combined with an architecture innovation, a parallel-sampling architecture is introduced in the next section to further improve the voltage efficiency (η_{vol}) of ADCs for the purpose of improving the ADC power efficiency for multi-carrier systems. Main advantage of this architecture is that it allows to further improve the signal range that can be processed linearly by an ADC (even beyond the linear signal range of the ADC sampling stage) compared to other voltage efficiency enhancement techniques discussed above. Hence, it enhances the SNR of the ADC for converting multi-carrier signals. It can also be combined with the other power reduction techniques discussed above towards power minimization of ADCs for multi-carrier systems.

3.5 A Parallel-Sampling ADC Architecture

The idea of using multiple ADCs to extend the dynamic range beyond the capability of currently available single ADCs was first presented in [30] for a radar system; the architecture diagram is shown in Fig. 3.12. In this architecture, each ADC is connected to the input through an amplifier which differs in gain from its

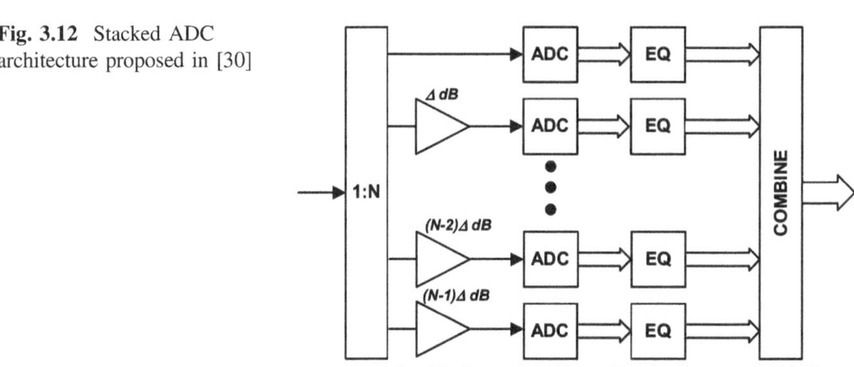

Fig. 3.12 Stacked ADC architecture proposed in [30]

adjacent ADC by Δ dB, and the outputs of the ADCs are combined after equalization to achieve a dynamic range (N-1)Δ dB greater than that of a single ADC. Validation of the concept was carried out on an ADC, named ‘stacked ADC’, built with commercial off-the-shelf devices (switches, stand alone ADCs, amplifiers, etc.). It demonstrated an improvement of its dynamic range over that of its sub-ADCs by approximately 30 dB. The author claimed that the overall SNDR of the ADC can not be improved compared to that of its sub-ADCs [30]. A similar ADC architecture using multiple ADCs in parallel was also proposed in [31] for the purpose of relaxing the ADC resolution requirements when used in a broadband communication system. As identified in this reference this concept can be used to improve both the dynamic range and SNDR of an ADC for converting broadband signals. In [32], another ADC architecture, composed of multiple ADCs in parallel to process the same input signal, was presented, but it is based on a different concept compared to the one mentioned above. The concept can be applied to ADCs to enhance their SNR by averaging the outputs of multiple ADCs. In this ADC architecture, the number of parallel sub-ADCs required quadrupled for every 6 dB increase in SNR.

Based on the same concept introduced in [30, 31] and with the ‘signal-aware’ design approach discussed in Chap. 2, the ADC architecture employing multiple lower dynamic range/resolution ADCs in parallel (the parallel-sampling ADC architecture) is analyzed and further developed for the purpose of improving the power efficiency of ADCs for multi-carrier signals by enabling a larger input signal range. In this section, the principle of the parallel-sampling ADC architecture for multi-carrier signals is presented; advantages of the architecture are analyzed and presented. Furthermore, in Sect. 3.6, run-time adaptation is introduced to the parallel-sampling ADC architecture to allow sub-ADCs sharing between different signal paths to further improve ADC power efficiency compared to previous works [30, 31].

3.5.1 Principle of the Parallel-Sampling ADC Architecture

A dual-path version of the parallel-sampling architecture is shown in Fig. 3.13a as an example, but the approach can be generalized easily to more paths. This ADC consists of two parallel sub-ADCs, each of them preceded by a range-scaling stage, and their outputs are combined by a signal reconstruction block. The principle of operation is shown in Fig. 3.13b: the front-end input signal is split into two signals that are scaled versions of each other after the range scaling stages. The signal in the main path has the same strength as the front-end input signal, while the signal in the auxiliary path is an attenuated version of the front-end input signal. These two signals are sampled by the sub-ADCs at the same time. Depending on the input signal level, one of them will be chosen to reconstruct the signal in the digital domain.

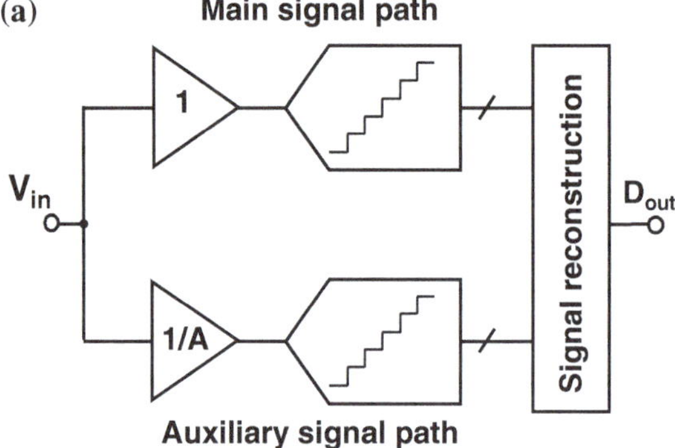

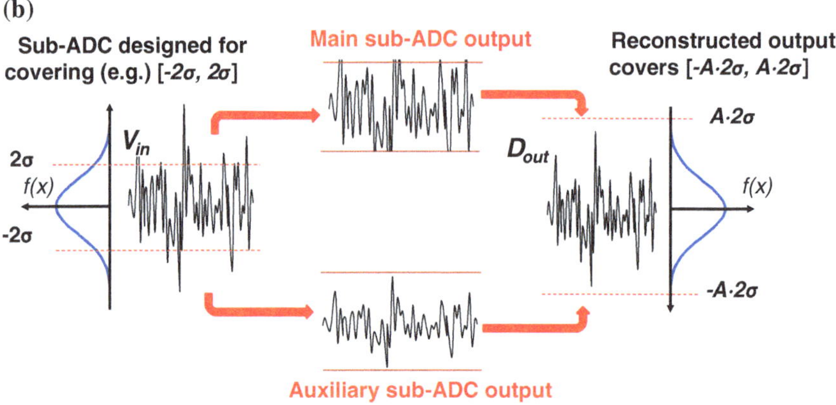

Fig. 3.13 **a** A dual-path version of the parallel-sampling architecture. **b** Principle of the parallel-sampling architecture for multi-carrier signals

The key idea of this architecture is that the signal in the main path is maximized to exploit the dynamic range of the sub-ADC more efficiently; hence the ADC has more resolution over the small amplitudes that have relatively much higher probability of occurrence due to the statistical multi-carrier signal amplitude properties. Large amplitudes that saturate the sub-ADC in the main path will be replaced in the digital domain by the samples from the auxiliary path. Since the sub-ADC in the auxiliary path quantizes an attenuated version of the ADC input signal, the probability of saturating the sub-ADC will be lower and the linearity better. During signal reconstruction, the auxiliary path provides coarsely quantized samples to replace the clipped or highly distorted large amplitude samples of the main path, hence avoiding excessive clipping noise and achieving good overall linearity. The signal reconstruction algorithm of the dual-path version is as follows:

$$D_{out}(n) = \begin{cases} D_{main}(n), & -V_{\max} < V_{in}(n \cdot T_s) < V_{\max} \\ A \cdot D_{auxiliary}(n), & else \end{cases} \tag{3.8}$$

where $D_{out}(n)$, $D_{main}(n)$, and $D_{auxiliary}(n)$ are the reconstructed signal, output of the main and auxiliary ADCs respectively, T_s is the sampling period, and $V_{\max}$ is the maximum input level of the sub-ADC.

3.5.2 Advantages of the Parallel-Sampling Architecture

As shown in Fig. 3.13, the input signal range of the parallel-sampling ADC and the digital word-length of the reconstructed signal can be larger than that of each sub-ADC. If the two sub-ADCs are chosen to have the same resolution, the sampled signal in the main path has a better SNR than the auxiliary one due to a larger input signal swing while the sampled auxiliary signal has a better SCDR. When the signal is reconstructed properly in the digital domain, the combination of the two sub-signals takes the advantages of both signal paths and offers a better SNCDR compared to that of each sub-ADC.

Figure 3.14 shows an example of mapping a Gaussian amplitude *pdf* with the input range of a single ADC and a dual path parallel-sampling ADC. The SNCDR of the ADCs can be expressed by:

$$SNCDR_{single\_ADC} = \frac{P_s}{P_n + P_c} = \frac{v_{s.RMS}^2}{v_{n.RMS}^2 + 2 \cdot \int_{L_{clip}}^{\infty} (x - L_{clip})^2 f(x) dx} \tag{3.9}$$

$$\begin{aligned} &SNCDR_{parallel\_ADC} \\ &= \frac{P_s}{F(R_{ADC1}) \cdot P_{n1} + F(R_{ADC2}) \cdot P_{n2} + P_c} \\ &= \frac{A^2 \cdot v_{s.RMS}^2}{\int_{-L_{clip}}^{L_{clip}} f(x) dx \cdot v_{n1.RMS}^2 + 2 \cdot \int_{A_{clip}}^{A \cdot L_{clip}} f(x) dx \cdot A^2 \cdot v_{n2.RMS}^2 + 2 \cdot \int_{A \cdot L_{clip}}^{\infty} (x - A \cdot L_{clip})^2 f(x) dx} \end{aligned} \tag{3.10}$$

where $F(x)$ is the distribution function of the input signal amplitudes; R_{ADC} is the signal range processed by the ADC; A is the attenuation factor of the signal in the auxiliary path; and $f(x)$ denotes the probability density function of the multi-carrier signal with zero mean and variance of $(L_{clip}/CR)^2$ as shown in Fig. 3.14.

From Eqs. (3.9) to (3.10), the optimal SNCDR of the ADC with and without parallel sampling can be found. In order to make the analysis more intuitive, the SNCDR of ideal ADCs (11 b and 12 b ADCs) without this technique is shown in

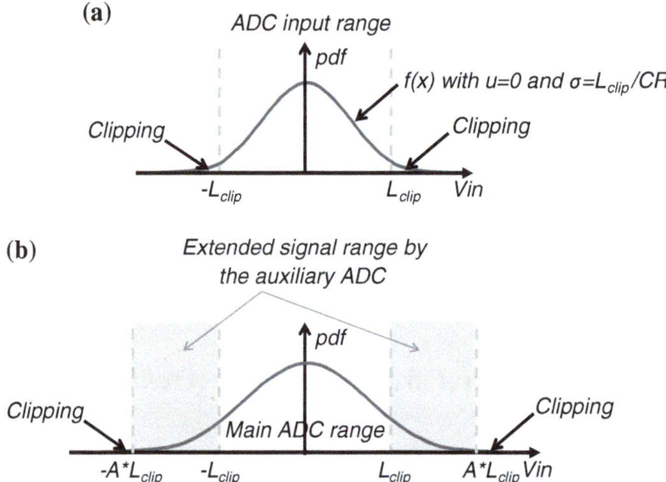

Fig. 3.14 Mapping of a multi-carrier signal amplitude probability distribution function to the input range of **a** a single ADC and **b** a dual path parallel-sampling ADC

Fig. 3.15 together with the SNCDR of the dual-path parallel-sampling ADCs (with two 11 b sub-ADCs) for different attenuation factor *A* (1.5–4 with a step of 0.5). By choosing a proper attenuation value for the auxiliary signal path, the SNCDR of the parallel-sampling ADC can be improved compared to that of its sub-ADCs. As shown in the plot, the dual path parallel-sampling ADC with *A* of 2 improves the SNCDR by about 5.5 dB compared to that of its sub-ADC (11 b) and is similar with that of a 12 b single ADC, while it consumes only 2 times the power of each sub-ADC. This observation is also valid for ADCs with arbitrary number of bits. When a single ADC is limited by thermal noise, its power consumption increases

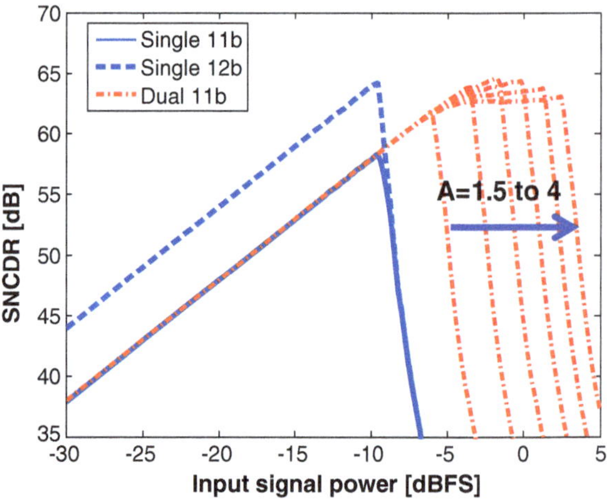

Fig. 3.15 SNCDR of the parallel-sampling ADC (with dual 11 b and different *A* values) and single 11 b and 12 b ADCs for converting multi-carrier signals

exponentially with conversion accuracy as discussed in Chap. 2. Therefore, for achieving a similar SNCDR requirement, the parallel-sampling ADC features higher power efficiency than the single ADC approach.

Analytical simulations using Eqs. (3.9) and (3.10) were performed to find the optimal power back-off values for an input signal having Gaussian amplitude *pdf*. The required power back-off of the parallel-sampling ADC for achieving a similar SNCDR is about 6 dB less compared to that of a single ADC with one extra bit as shown in Fig. 3.16. The SNCDR improvement of the parallel-sampling ADC over its sub-ADC is achieved by the ability to process a larger input signal swing without causing excessive clipping distortion as can be observed from the output spectrum shown in Fig. 3.17.

It was discussed in Sect. 3.4 that processing a large signal swing is normally constrained by the linearity of the input sampling stage of an ADC. In the parallel-sampling architecture, this large input signal swing does not need to pass through the sub-ADCs. The sub-ADC in the main path is only required to convert the small amplitude values linearly. Compressing and clipping the large signal amplitudes in the main path do not affect the final linearity of the reconstructed output signal. For the sub-ADC in the auxiliary path, the range-scaling block in front of the sub-ADC attenuates the large input signal and translates it into the linear signal range of the sub-ADC.

For A of 2, the required linear input signal range of the sub-ADC can be only half of the ADC's front-end input signal range. Therefore the input sampling stages of the sub-ADCs can be implemented with thin-oxide transistors with short channel length which is beneficial for higher sampling rates. Compared to previous work [25] which relies on using thick-oxide I/O transistors with high supply voltage to

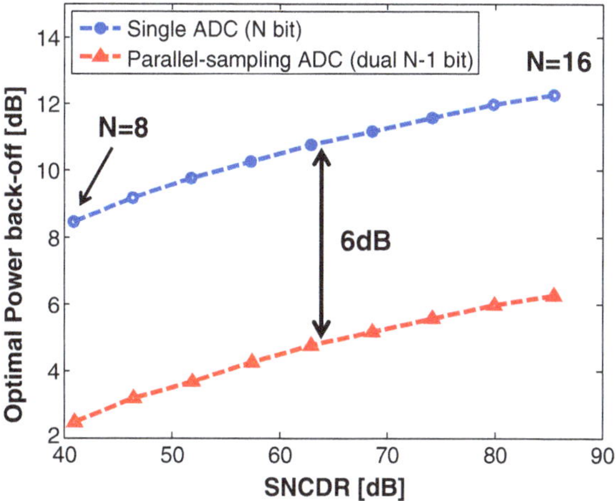

Fig. 3.16 Comparison of the optimal power back-off values of the dual path parallel-sampling ADC and the single ADC for getting a similar SNCDR

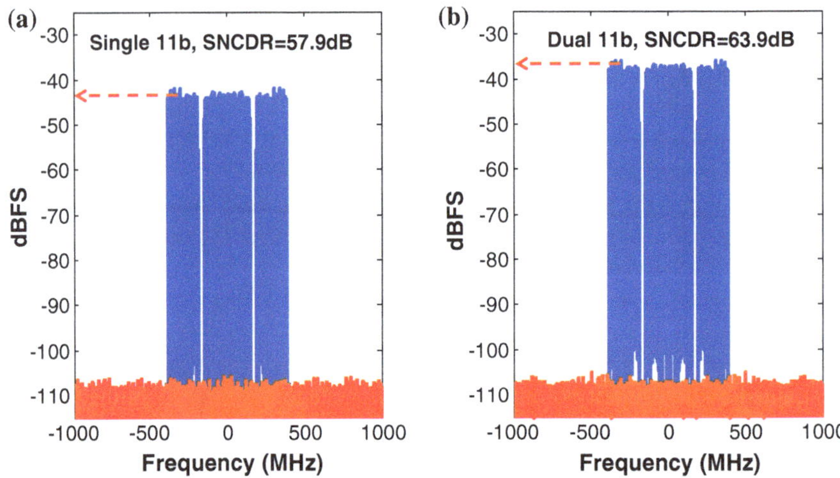

Fig. 3.17 Spectrum plots of the outputs of **a** a single 11 b ADC and **b** a dual path parallel-sampling ADC with two 11 b sub-ADCs, with their input signal at optimal power back-off for maximum SNCDR

achieve a large input signal range, this architecture has an advantage in processing a large input signal swing without sacrificing speed (thick-oxide option can be avoided). This parallel-sampling architecture is especially useful in an advanced CMOS process that allows simultaneously achieving both high speed and larger signal range.

In summary, this ADC architecture enables a larger input signal range without causing excessive distortion and allows using ADCs with lower resolution to achieve a similar SNCDR as a single ADC with higher resolution. An enlarged input signal range leading to higher voltage efficiency [22, 23, 25] is the key to improve the power efficiency, as a desired SNR can be achieved with a much smaller sampling capacitor. For example, increasing the signal range by 2 times allows reducing the total sampling capacitor by 4 times for getting the same SNR. The reduced sampling capacitor size also brings advantage in improving the signal bandwidth and sampling rate of the ADC. For thermal noise limited ADCs, one extra bit corresponds to a 4 times increase in power and area; allowing two ADCs with one bit less in resolution to fulfill a similar system requirement, saves up to 50 % in power and area.

There are still two observations to be made. Firstly, while this architecture is intended to improve the ADC power efficiency for multi-carrier signals that have a 'bell-shaped' amplitude probability distribution function (e.g. Gaussian amplitude *pdf*) for a desired SNCDR, the principle of using multiple ADCs with different ranges can also be applied to signals with other *pdf* shapes. However, the advantage doesn't apply to signals with flat amplitude probability distribution. Secondly, the driver of the parallel-sampling ADC has to deliver a larger output signal swing which can be a limitation of this architecture. However, the reduction of the sampling capacitor makes the ADC easier to drive and doesn't affect the overall power efficiency [15, 22, 33] and the required large output signal swing of the ADC driver can be achieved by using the mixed-supply-voltage approach without compromising the speed as proposed in [26, 27].

3.5.3 Impact of Mismatch Between Signal Paths

As the reconstructed signal requires a proper combination of multiple parallel sub-ADC outputs, the ADC performance can be degraded by mismatch between different signal paths. There are several sources of mismatch in a parallel-sampling ADC (a dual path version is shown in Fig. 3.18 as an example) which are caused by the different offset os_i and timing skew Δt_i in each signal path, as well as the difference of the attenuation factor A in the analog circuit paths and its corresponding correction factor A' in the digital circuit.

Equation (3.8) in the previous section can be rewritten to include these errors, as shown in Eq. (3.11). It can also be expanded to include other non-idealities.

$$D_{out}(n) = \begin{cases} Q(V_{in}(n \cdot T_s + \Delta t_1) + os_1), & -V_{max} < V_{in}(n \cdot T_s) < V_{max} \\ A' \cdot Q(1/A \cdot V_{in}(n \cdot T_s + \Delta t_2) + os_2) & -A \cdot V_{max} < V_{in}(n \cdot T_s) < A \cdot V_{max} \end{cases} \tag{3.11}$$

where $Q(x)$ denotes the quantization function, os_i is the offset voltage, Δt_i is the timing mismatch, and A' is the digital gain to restore the auxiliary ADC output.

These sources of mismatch resemble that of the well-known time-interleaved ADC architecture [34], but their impact on the ADCs performance differs for the two architectures. The impact of the mismatch on the time-interleaved ADC performance has been analyzed extensively [35–37]. In the following section, the differences in impact of the errors on the performance of these two architectures are analyzed.

Firstly, the sub-ADCs in the parallel-sampling ADC are sampled synchronously instead of in a time-interleaved fashion. The errors due to mismatch are only

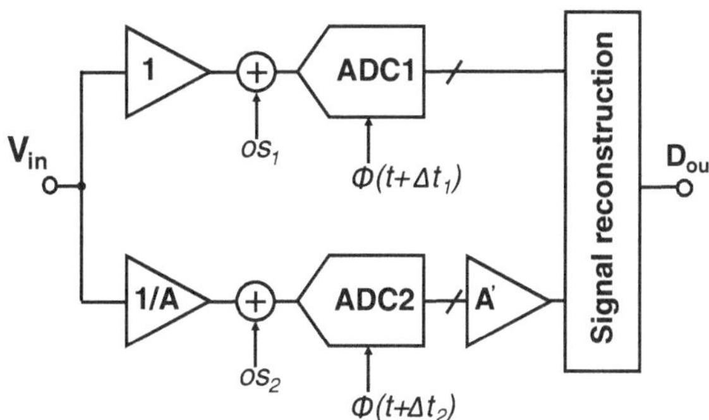

Fig. 3.18 Gain, offset, and timing mismatches in a dual path version parallel-sampling ADC

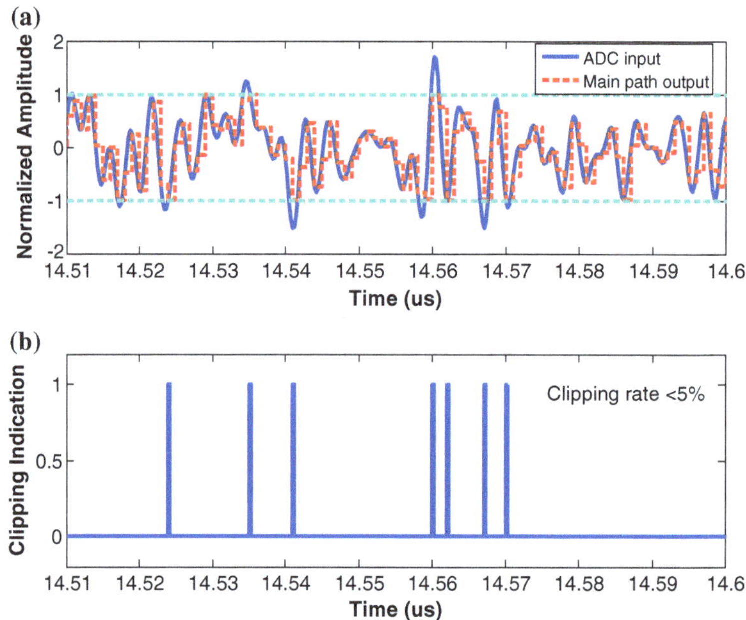

Fig. 3.19 **a** Signal waveform of the parallel-sampling ADC (sub-ADC input range between −1 and 1); **b** clipping indication of the main signal path

introduced when the samples from the main sub-ADC need to be replaced with the corresponding samples from the auxiliary sub-ADC. Due to the amplitude property of the multi-carrier signal, clipping events don't appear in a repetitive pattern, as shown in Fig. 3.19. Therefore, the mismatch induced errors do not affect the reconstructed signal in a repetitive manner while they do so in a time-interleaved ADC. Secondly, the probability of signal amplitudes that clip the main sub-ADC is low for multi-carrier signals with a bell-shaped amplitude *pdf*. When the input signal power of the parallel-sampling ADC is backed off for the maximum SNCDR, the number of samples from the auxiliary sub-ADC that are used to replace the samples from the main sub-ADC is small compared to the total number of samples of the reconstructed signal: less than 5 % for a dual 11 b parallel-sampling ADCs with $A = 2$ as shown in Fig. 3.19. For a two-path time-interleaving ADC, every other sample introduces mismatch errors (corresponding to 50 % of the total number of samples). As a result, the error power due to mismatch is much smaller than that of time-interleaved ADCs considering similar matching performance. In Fig. 3.20, the relationship of the maximum SNCDR with different offset and gain mismatches are shown. It confirms that the performance of the parallel-sampling ADC architecture is much less sensitive compared to that of the time-interleaving ADC according to the analysis above.

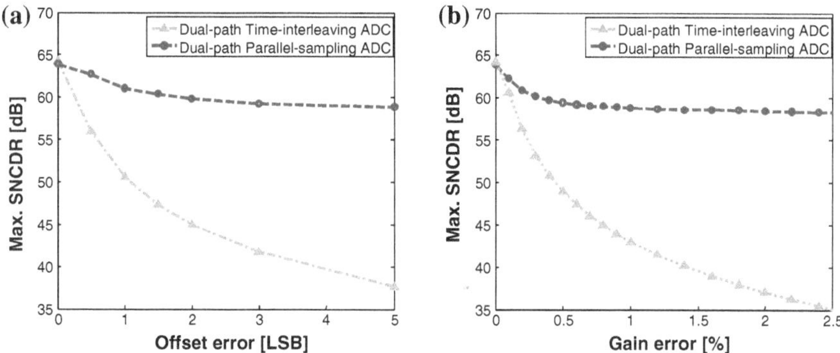

Fig. 3.20 Impact of mismatch between sub-ADCs on the SNCDR of the parallel sampling ADC (*dual* 11 b *ADC with A* = 2) and a time-interleaving ADC (*dual* 11 b) for a wideband multi-carrier signal due to **a** offset and **b** gain mismatch

Errors caused by mismatch between different signal paths can be minimized by design or calibration. The ways to calibrate the mismatch errors of the parallel-sampling architecture are similar to that of the time-interleaving ADC architecture [38–40]. System simulation can be performed to derive the matching requirements for achieving a certain performance.

3.6 Implementation Options

This section presents four implementation options for the parallel-sampling ADC architecture. Theoretically, a higher SNCDR can be achieved with a larger number of parallel-sampling paths and a larger input signal swing. In a practical implementation, the maximum allowed input signal swing on chip is constrained by the linear range of the circuits and further by the reliability of the components in a specific technology.

The first implementation option of the dual-path version is shown in Fig. 3.21a. It consists of a range-scaling stage, two sub-ADCs (the front-end sampler of the ADC is shown separately), a signal reconstruction block and an out-of-range detection block which is an optional block to detect whether the instantaneous signal amplitude is within the linear signal range of the main path or not. The out-of-range signal detection function can also be embedded in one of the sub-ADCs in a practical implementation.

The second implementation option is shown in Fig. 3.21b. By introducing run-time adaptation to the architecture, part of the ADC function can be shared by both signal paths to further reduce the power consumption and area. The run-time adaptation is realized by introducing an analog multiplexer between the front-end samplers and the sub-ADC and controlling it by a dedicated out-of-range detection block. Therefore, instead of using two independent sub-ADCs (the first

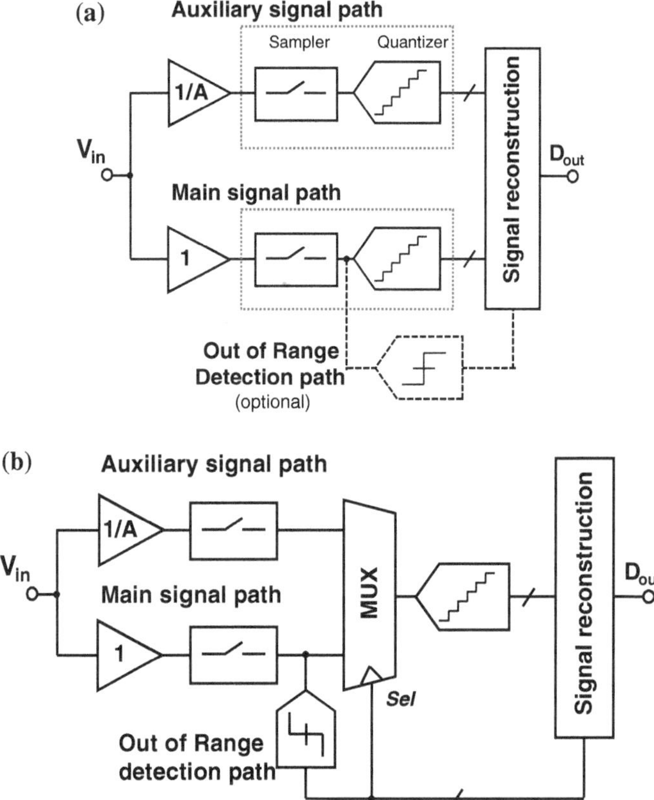

Fig. 3.21 Two implementation options of the dual-path parallel-sampling architecture: **a** with two separated sub-ADCs and **b** with run-time adaptation to share the sub-ADC between two signal paths

implementation option), only one sub-ADC is needed by sharing it between two signal paths. This improves the power and area efficiency of this architecture while at the cost of extra design complexity (to deal with the non-idealities due to the additional analog multiplexer in the critical signal paths).

The third implementation option is shown in Fig. 3.22. The input signal range covered by the auxiliary path overlaps that of the main signal path in the implementation one and two. It is possible to reduce the dynamic range requirement of the sub-ADCs without sacrificing performance by reducing this overlapping. A subtraction block can therefore be introduced before the front-end sampler of the auxiliary signal paths to reduce the dynamic range requirement of the sub-ADCs in the auxiliary signal paths.

The fourth implementation option is shown in Fig. 3.23. By introducing adaptation to both the attenuation factor A and the strength of the input signal V_{in}, the input signal range covered by the main and auxiliary signal paths can be adapted

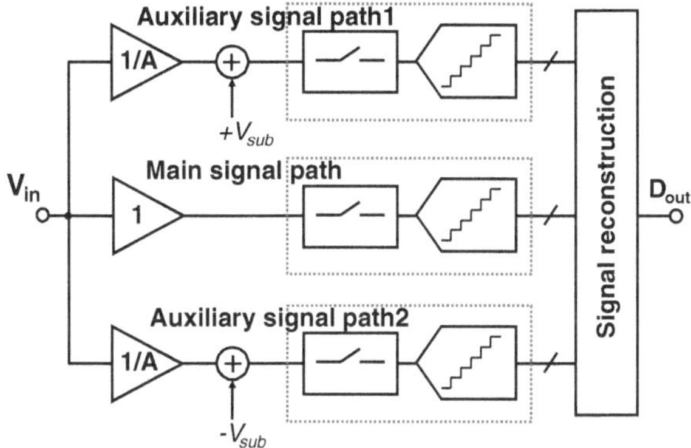

Fig. 3.22 Implementation of the parallel-sampling architecture with subtraction to reduce the dynamic range requirement of the sub-ADCs in the auxiliary signal paths

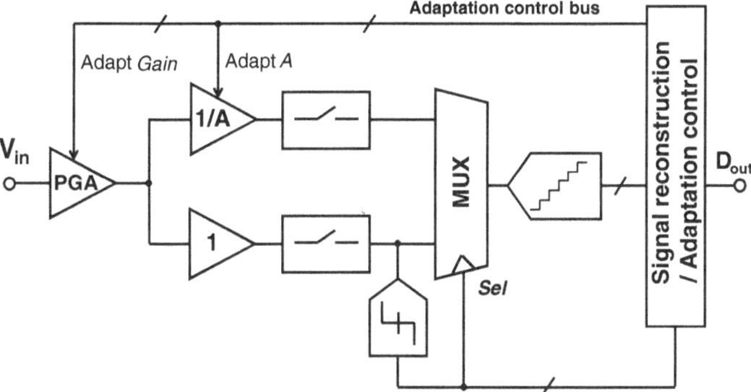

Fig. 3.23 Implementations of the parallel-sampling architecture with adaptation control of the attenuation factor *A* and input signal strength

according to the actual statistical properties of the input signal at the cost of additional design complexity and hardware. In this way, the ADC takes advantage of both a priori and a posteriori information of the signal.

3.7 Conclusions

In this chapter, broadband multi-carrier transmission was briefly introduced. Statistical amplitude properties of the multi-carrier signal and the ADC's dynamic range requirement for multi-carrier systems were analyzed. The tradeoff between power consumption and SNR of thermal noise limited ADCs was discussed. Power reduction techniques at circuit and architecture level for thermal noise limited

ADCs in advanced CMOS technologies were studied and summarized. It was shown that improving the voltage efficiency (η_{vol}) of thermal noise limited ADCs is an effective way to enhance the ADC power efficiency for a desired SNR. By exploiting signal properties (the statistical signal amplitude properties) and architecture innovation, a parallel-sampling architecture was introduced to further enhance the η_{vol} of ADCs for the purpose of improving the ADC power efficiency for multi-carrier systems. Main advantage of this architecture is that it allows to further improve the signal range that can be processed linearly by an ADC (even beyond the linearity signal range of the ADC sampling stage) compared to other voltage efficiency enhancement techniques. This architecture can also be combined with the other power reduction techniques discussed earlier towards power minimization of ADCs for multi-carrier systems. Four implementation options of the parallel-sampling ADC architecture were proposed, which includes an implementation option with run-time adaptation to allow sub-ADCs sharing between different signal paths to further improve ADC power efficiency compared to previous works [30, 31]. From analytical analysis and simulation of the parallel-sampling ADC architecture, we conclude that ADCs employing this architecture have the advantage of reducing power consumption and area for multi-carrier signals to achieve a desired SNCDR.

This chapter emphasized on the concept and architectural study. The next chapter (Chap. 4) will focus on circuit implementations of the parallel-sampling architecture. A parallel-sampling first stage of a 12 b pipeline ADC and a parallel-sampling frontend stage of a 4 GS/s 11 b time-interleaving SAR ADC will be presented. Furthermore, an IC implementation of a parallel-sampling ADC (with two 1 GS/s 11 b sub-ADC) will be presented to verify this concept.

References

1. ISSCC 2013: Trends. 2014. [Online]. http://www.isscc.org/doc/2013/2013_Trends.pdf. Accessed 09 Jun 2014.
2. Shannon, C.E. 1998. Communication in the presence of noise. *Proceedings of the IEEE* 86(2): 447–457.
3. Pierce, J.R. 1980. *An introduction to information theory: Symbols, signals and noise*, Subsequent ed. Mineola, NY: Dover Publications.
4. Bingham, J.A.C. 1990. Multicarrier modulation for data transmission: An idea whose time has come. *IEEE Communications Magazine* 28(5): 5–14.
5. Pun, M., M. Morelli, and C.C.J. Kuo. 2007. *Multi-carrier techniques for broadband wireless communications: A signal processing perspectives*. London: Singapore; Hackensack, NJ: Imperial College Press.
6. El-Absi, M., and M. Banat. 2011. *Peak-to-average power ratio reduction in OFDM systems: Toward more efficient and reliable wireless systems*. Germany: LAP LAMBERT Academic Publishing.
7. IEEE Standard for Information Technology: Wireless LAN Medium Access Control (MAC) and Physical Layer (PHY) Specifications. 2013. *IEEE Std 802.11ac-2013*, 1–425.
8. DOCSIS 3.0 Physical Layer Interface Specification. 2014. [Online]. http://www.cablelabs.com/specification/docsis-3-0-physical-layer-interface-specification/. Accessed 09 Jun 2014.

9. Gustavsson, M., J.J. Wikner, and N. Tan. 2000. *CMOS data converters for communications*, 2000th ed. Boston: Springer.
10. Feller, W. 1945. The fundamental limit theorems in probability. *Bulletin of the American Mathematical Society* 51(11): 800–832.
11. Clipping in Multi-Carrier Systems. 2004. *Multi-carrier digital communications*, 69–98. US: Springer.
12. Chiu, Y., B. Nikolic, and P.R. Gray. 2005. Scaling of analog-to-digital converters into ultra-deep-submicron CMOS. In *Custom Integrated Circuits Conference, 2005. Proceedings of the IEEE 2005*, 375–382.
13. Bult, K. 2000. Analog design in deep sub-micron CMOS. In *Solid-State Circuits Conference, 2000. ESSCIRC '00. Proceedings of the 26rd European*, 126–132.
14. Annema, A.-J., B. Nauta, R. van Langevelde, and H. Tuinhout. 2005. Analog circuits in ultra-deep-submicron CMOS. *IEEE Journal of Solid-State Circuits* 40(1): 132–143.
15. Murmann, B. 2010. Trends in low-power, digitally assisted A/D conversion. *IEICE Transactions on Electronics* E93-C(6): 718–729.
16. Razavi, B. 2000. *Design of analog CMOS integrated circuits*. New Delhi: McGraw-Hill Education.
17. Sansen, W. 2012. Power minimization in ADC design. In *Analog circuit design*, ed. M. Steyaert, A. van Roermund, and A. Baschirotto, 3–18. The Netherlands: Springer.
18. Murmann, B. 2012. Low-power pipelined A/D conversion. In *Analog circuit design*, ed. M. Steyaert, A. van Roermund, and A. Baschirotto, 19–38. The Netherlands: Springer.
19. Horowitz, M., E. Alon, D. Patil, S. Naffziger, R. Kumar, and K. Bernstein. 2005. Scaling, power, and the future of CMOS. In *Electron Devices Meeting, 2005. IEDM Technical Digest. IEEE International*, 7–15.
20. Taiwan Semiconductor Manufacturing Company Limited. 2014. [Online]. http://www.tsmc.com/english/dedicatedFoundry/technology/28nm.htm. Accessed 14 Jun 2014.
21. Abo, A.M., and P.R. Gray. 1999. A 1.5-V, 10-bit, 14.3-MS/s CMOS pipeline analog-to-digital converter. *IEEE Journal of Solid-State Circuits* 34(5): 599–606.
22. Stroeble, O., V. Dias, and C. Schwoerer. 2004. An 80 MHz 10 b pipeline ADC with dynamic range doubling and dynamic reference selection. In *Solid-State Circuits Conference, 2004. Digest of Technical Papers. ISSCC. 2004 IEEE International*, 462–539, Vol. 1.
23. Van de Vel, H., B.A.J. Buter, H. Van Der Ploeg, M. Vertregt, G.J.G.M. Geelen, and E.J.F. Paulus. 2009. A 1.2-V 250-mW 14-b 100-MS/s digitally calibrated pipeline ADC in 90-nm CMOS. *IEEE Journal of Solid-State Circuits* 44(4): 1047–1056.
24. Mak, P.-I., and R.P. Martins. 2010. High-/mixed-voltage rf and analog CMOS circuits come of age. *IEEE Circuits and Systems Magazine* 10(4): 27–39 (Fourthquarter).
25. Choi, H.-C., J.-W. Kim, S.-M. Yoo, K.-J. Lee, T.-H. Oh, M.-J. Seo, and J.-W. Kim. 2006. A 15 mW 0.2 mm/sup 2/ 50 MS/s ADC with wide input range. In *Solid-State Circuits Conference, 2006. ISSCC 2006. Digest of Technical Papers. IEEE International*, 842–851.
26. Wu, J., A. Chou, C.-H. Yang, Y. Ding, Y.-J. Ko, S.-T. Lin, W. Liu, C.-M. Hsiao, M.-H. Hsieh, C.-C. Huang, J.-J. Hung, K. Y. Kim, M. Le, T. Li, W.-T. Shih, A. Shrivastava, Y.-C. Yang, C.-Y. Chen, and H.-S. Huang. 2013. A 5.4 GS/s 12 b 500 mW pipeline ADC in 28 nm CMOS. In *2013 Symposium on VLSI Circuits (VLSIC)*, C92–C93.
27. Janssen, E., K. Doris, A. Zanikopoulos, A. Murroni, G. van der Weide, Y. Lin, L. Alvado, F. Darthenay, and Y. Fregeais. 2013. An 11 b 3.6 GS/s time-interleaved SAR ADC in 65 nm CMOS. In *Solid-State Circuits Conference Digest of Technical Papers (ISSCC), 2013 IEEE International*, 464–465.
28. Weltin-Wu, C., and Y. Tsividis. 2013. An event-driven clockless level-crossing ADC with signal-dependent adaptive resolution. *IEEE Journal of Solid-State Circuits* 48(9): 2180–2190.
29. Wakin, M., S. Becker, E. Nakamura, M. Grant, E. Sovero, D. Ching, J. Yoo, J. Romberg, A. Emami-Neyestanak, and E. Candes. 2012. A nonuniform sampler for wideband spectrally-sparse environments. *IEEE Journal on Emerging and Selected Topics in Circuits and Systems* 2(3): 516–529.

30. Gregers-Hansen, V., S.M. Brockett, and P.E. Cahill. 2001. A stacked A-to-D converter for increased radar signal processor dynamic range. In *Proceedings of the 2001 IEEE Radar Conference, 2001*, 169–174.
31. Doris, K. 2008. Data processing device comprising adc unit. WO2008072130 A1.
32. Reeder, Rob, Mark Looney, and Jim Hand. 2005. Pushing the state of the art with multichannel A/D converters. *Analog Dialogue* 39–2: 7–10.
33. Lin, Y., K. Doris, H. Hegt, and A.H.M. Van Roermund. 2012. An 11 b pipeline ADC with parallel-sampling technique for converting multi-carrier signals. *IEEE Transactions on Circuits and Systems I: Regular Papers* 59(5): 906–914.
34. Black, W.C., and D. Hodges. 1980. Time interleaved converter arrays. *IEEE Journal of Solid-State Circuits* 15(6): 1022–1029.
35. Kurosawa, N., H. Kobayashi, K. Maruyama, H. Sugawara, and K. Kobayashi. 2001. Explicit analysis of channel mismatch effects in time-interleaved ADC systems. *IEEE Transactions on Circuits and Systems I: Fundamental Theory and Applications* 48(3): 261–271.
36. El-Chammas, M., and B. Murmann. 2012. Time-interleaved ADCs. In *Background calibration of time-interleaved data converters*, 5–30. New York: Springer.
37. Razavi, B. 2013. Design considerations for interleaved ADCs. *IEEE Journal of Solid-State Circuits* 48(8): 1806–1817.
38. Fu, D., K.C. Dyer, S.H. Lewis, and P.J. Hurst. 1998. A digital background calibration technique for time-interleaved analog-to-digital converters. *IEEE Journal of Solid-State Circuits* 33(12): 1904–1911.
39. Doris, K., E. Janssen, C. Nani, A. Zanikopoulos, and G. Van Der Weide. 2011. A 480 mW 2.6 GS/s 10 b time-interleaved ADC with 48.5 dB SNDR up to Nyquist in 65 nm CMOS. *IEEE Journal of Solid-State Circuits* 46(12): 2821–2833.
40. Dusan Stepanovic. 2012. *Calibration techniques for time-interleaved SAR A/D converters*. Ph.D dissertation, EECS Department, University of California, Berkeley.

Chapter 4
Implementations of the Parallel-Sampling ADC Architecture

Abstract This chapter describes circuit implementations of the parallel-sampling ADC architecture presented in Chap. 3. The parallel-sampling architecture is applied to two ADC architectures (a pipeline and a time-interleaving SAR ADC architecture), which are suitable for designing high-speed and medium-to-high resolution ADCs, to improve the ADC power efficiency for multi-carrier signals. Section 4.1 describes the architecture and operation of a 200 MS/s 12-b switched-capacitor pipeline ADC with a parallel-sampling first stage, which is suitable for broadband multi-carrier receivers for wireless standards such as LTE-advanced and the emerging generation of Wi-Fi (IEEE802.11ac) . A circuit implementation of the parallel-sampling first stage of the pipeline ADC is presented and simulation results are given. Section 4.2 presents the architecture and operation of a 4 GS/s 11 b time-interleaved ADC with a parallel-sampling frontend stage, which targets wideband direct sampling receivers for DOCSIS 3.0 cable modems . Circuit implementation and simulation of the 4 GS/s parallel-sampling frontend stage are given. Due to the complexity of implementing the proposed 4 GS/s ADC on chip, a two-step design approach was adopted. In Sect. 4.3, a prototype IC of an 11 b 1 GS/s ADC with a parallel sampling architecture is presented, which serves as a first step to validate the parallel-sampling ADC concept and the performance of the high-speed parallel-sampling frontend and detection circuits. In future work, the frontend stage of the IC can be interleaved by four times to achieve the aggregate sample rate of 4 GHz of the proposed ADC discussed in Sect. 4.2. Conclusions of this chapter are drawn in Sect. 4.4.

Y. Lin et al., *Power-Efficient High-Speed Parallel-Sampling ADCs for Broadband Multi-carrier Systems*, Analog Circuits and Signal Processing,
DOI 10.1007/978-3-319-17680-2_4

4.1 Parallel-Sampling Architecture Applied to a Pipeline ADC

This section describes the architecture and circuit design of a 200 MS/s 12-b switched-capacitor pipeline ADC with a parallel-sampling first stage [1]. The proposed parallel-sampling first stage can be applied to pipeline ADCs with different accuracy and speed specifications for multi-carrier signals.

4.1.1 Pipeline ADCs Architecture

In broadband multi-carrier receivers for wireless standards such as LTE-advanced and the emerging generation of Wi-Fi (IEEE802.11ac) which use OFDM modulation techniques [2, 3], ADCs with resolutions of 10-b or higher and sampling rates of over a hundred MHz are required. The pipeline ADC is a suitable architecture for achieving such specifications with good power efficiency [4]. A general structure of the pipeline ADC architecture is shown in Fig. 4.1 [5]. The pipeline ADC consists of a number of low resolution stages and a digital correction and encoding block. These stages can operate concurrently by alternating their operations between sampling the input/residue signal from the previous stage and producing a residue signal for the next stage [5]. Each stage resolves a certain number of bits and generates a residue signal that is digitized by the succeeding stages. The digital bits of all the stages are time aligned and combined to form the output. A typical

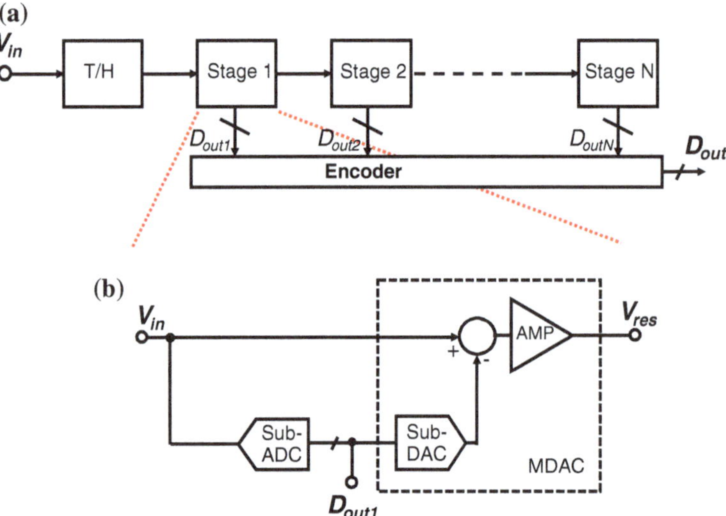

Fig. 4.1 **a** General pipelined ADC architecture; **b** typical pipeline stage inner structure

pipeline stage is shown in Fig. 4.1b. It consists of a sub-ADC, a sub-DAC, a subtractor and a residue gain amplifier. It is common practice to implement the sub-DAC together with the subtraction block and the amplifier as a single block, called Multiplying-DAC (MDAC). The last stage needs only a sub-ADC. The sub-ADC of a pipeline ADC is normally a flash ADC. Sometimes a dedicated track-and-hold (T/H) stage is placed at the input to avoid timing skew between signal paths to the subtractor and the sub-ADC.

The main advantage of the pipeline architecture is that thanks to stage pipelining, the maximum sampling frequency of the converter is determined only by the time period of a conversion cycle of a single stage. The propagation time through the cascade of pipeline stages results only in latency, meaning the time delay between an analog input and its digital representation. Depending on the application, latency can cause problems, for example in case the ADC is used in a feedback path of a system. One of the critical design choices of a pipeline ADC is the number of bits that is produced by each stage, which can vary in each stage in a pipeline ADC. By making a correct distribution of the number of bits over the stages in the pipeline ADC, the overall speed, accuracy, power consumption and chip area can be optimized [6, 7].

4.1.2 A Parallel-Sampling First Stage for a Pipeline ADC

As discussed in Chap. 3, enabling the ADC to process a large signal range (hence to provide a higher voltage efficiency) is an effective way to improve the power efficiency of a thermal noise limited ADC. Enlarging the input signal range by two times allows a four times reduction of the size of sampling capacitors while still getting a similar SNR, which results in a similar power reduction for amplifiers that need to drive them. In a pipeline ADC, the maximum input signal swing that can be handled linearly is normally constrained by the sampling circuit or the amplifier's output stage, as illustrated in Fig. 4.2 which shows two typical circuit building blocks of a pipeline stage and their transfer function. The linear signal range of the sampling circuit is limited by the input dependent on-resistance of the MOS switch, while the linear output signal range of an amplifier can be only a fraction of the supply voltage due to the voltage headroom limitation (e.g. 0.2–0.4 for a telescopic cascode amplifier.). Bootstrapping technique can be applied to the sampling switches to achieve a larger linear signal range, but it is challenging to achieve both good linearity (e.g. above 70 dB) and high speed operation due to various circuit non-idealities [8, 9]. The parallel-sampling architecture presented in Chap. 3 can be applied to the pipeline ADC to exceed the linear signal range limitation set by its sampling circuit and the amplifier's output stage. The first stage of a pipeline ADC is the most critical one in terms of speed and noise performance, hence dominating the power consumption of the ADC (e.g. consuming 30–40 % of the total power consumption) [6, 7, 10, 11]. The power consumption of succeeding stages is decreased strongly by stage scaling [6, 7, 10]. Therefore, the parallel-sampling

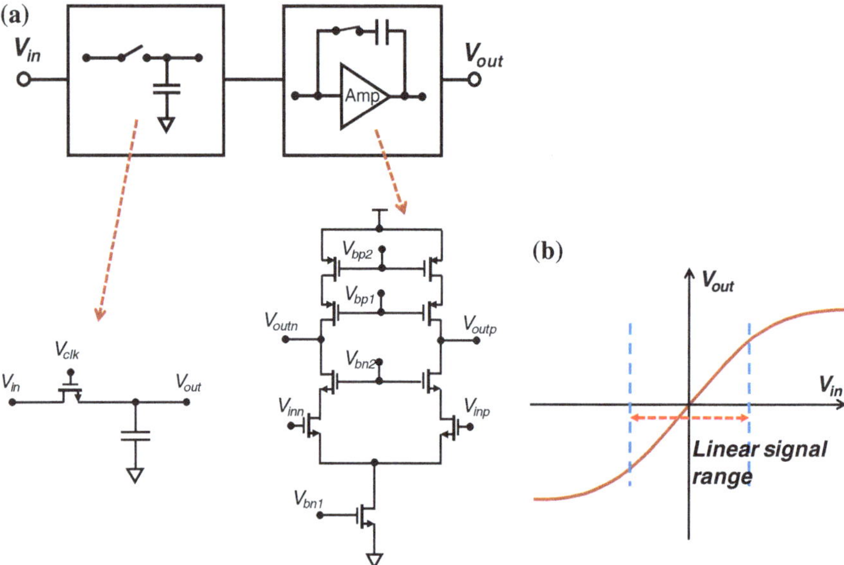

Fig. 4.2 **a** Tow typical circuit building blocks of a pipeline stage: a sampling circuit and a reside amplifier (a single stage class-A amplifier for example); **b** signal transfer function illustrating the linear signal range

architecture is only applied to the first stage of a 12-b switched-capacitor pipeline ADC for the purpose of improving the ADC power efficiency for multi-carrier signals. It can also be used in combination with conventional design techniques to improve further the linear input signal range of an ADC, and hence reduce the required sampling capacitor size and the ADC power consumption for getting a certain SNR.

Figure 4.3a shows the architecture of the proposed ADC. It consists of a parallel-sampling first stage and a backend ADC implemented with conventional pipeline stages. This ADC operates without a dedicated frontend T/H stage to reduce the power consumption [12, 13]. Figure 4.3b shows the block diagram of the parallel-sampling first stage of the pipeline ADC. In this stage, there are three paths for the input signal: the main signal path, the auxiliary signal path and the detection path. The main and auxiliary signal paths each consist of a signal scaling block and a passive sampling network (T/H) that are multiplexed by a signal paths selection block (MUX) to a subtraction block and then to the amplification block (AMP). The detection path contains a Flash sub-ADC (with additional comparison levels for out-of-range detection). Compared to the conventional pipeline stage as shown in Fig. 4.1, instead of maintaining the same input and output signal range in the first stage, adaptive signal paths selection through a MUX is introduced to decouple the choice of the stage's input and output signal swing. This adaptive signal paths selection is input signal level dependent and works on a sample-by-sample basis. When the instantaneous signal amplitude at the ADC input is within the desired

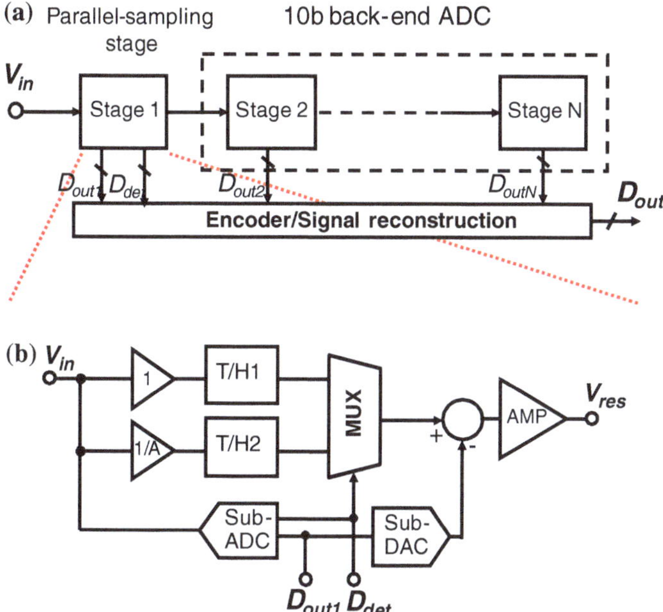

Fig. 4.3 **a** The architecture of the proposed pipeline ADC; **b** block diagram of the first pipelined stage with parallel sampling technique

linear range of the T/H or the amplifier, the sampled signal in the main signal path is selected for further conversion; when the instantaneous signal amplitude at the ADC input is beyond the desired linear range of the T/H and the amplifier, the sampled signal in the auxiliary signal path, which is an attenuated version of the sampled signal in the main signal path, is selected for further conversion (the change in signal gain is compensated in the digital encoder). In this way, the effective linear input signal range of the ADC can be enlarged by a factor of A, while clipping of instantaneous signal amplitudes with large values can be avoided or reduced significantly. This proposed ADC architecture is similar to the implementation option shown in Fig. 3.21 in Chap. 3.

Advantages brought by this parallel-sampling stage are as follows. Firstly, the ‘scaling’ of the signal range and adaptive signal paths selection relax the linearity requirement of the T/Hs and the residue amplifier for a large input signal. Secondly, the reduction of the capacitor size for a desired SNR leads to a substantial power reduction of the amplifier used for residue generation. Thirdly, with the adaptive signal paths selection, the back-end stages of this pipeline ADC (the 10 b back-end ADC shown in Fig. 4.3a) are shared by two signal paths instead of using two ADCs in parallel as that of the first implementation option shown in Fig. 3.21 in Chap. 3, which improves power and area efficiency. The residue amplifier in the first stage is also shared by the two signal paths, which improve further the power efficiency, since most of the power in a pipeline stage is consumed by the amplifiers used for residue generation (all the residue amplifiers together can consume more than half of the overall pipeline ADC power [6, 10]). Fourthly, the additional two comparators in the Sub-ADC needed for signal-range detection consumes much smaller power compared to the residue amplifier as their accuracy requirements can be relaxed by redundancy and error correction [5, 6].

In the following sections, implementation, operation and simulation of the parallel-sampling first stage will be presented. This stage is designed and simulated using CMOS 65 nm technology with only thin-oxide transistors and a single 1.2 V supply voltage. The target input referred thermal noise power is 70 dBFS (corresponds to a conventional ADC with about 11.3 ENOB) and the sampling speed is 200 MHz.

4.1.3 Implementation and Operation of the First Stage

The proposed parallel-sampling first stage operates as a 2.5 b stage (with 0.5 b redundancy [5, 6]) when the main path is selected and a single bit stage when the auxiliary path is selected. The 2.5 b for the main signal path (6 comparison levels) is chosen for a good trade-off between power and speed [5–7, 10], and with a signal gain of 4 to relax the requirements on the backend stages. The resolution in the auxiliary path is chosen to be 1 b for simplicity, since the probability of utilizing the output of this path is much lower than that of the main path.

A schematic representation of the first stage is shown in Fig. 4.4, as well as its timing diagram. The actual circuit implementation is fully differential. The linear signal range of the T/Hs and the output of residue amplifier are designed to be 0.8 V peak-to-peak differential (V_{ppd}). The parallel sampling architecture enables the ADC to have an effective input signal range of $2V_{ppd}$ which improves the voltage efficiency by a factor of 2.5 compared to that of the output stage of the amplifier with a single low supply voltage of 1.2 V.

The signal scaling blocks on the left are implemented by polysilicon resistors. The resistive divider in the auxiliary path attenuates the input signal swing by a factor of 2.5 (an attenuation factor chosen for achieving a close to optimal SNCDR found by system level simulations as explained in Chap. 3 and taking into account implementation complexity), while the resistor circuit in the main signal path keeps the signal un-attenuated. The resistive range scaling block also provides a 50 Ω input termination. The unit resistors are sized to have an intrinsic matching according to the design target of 12 b accuracy and the timing and bandwidth mismatches between two signal paths are minimized by design (devices sizing and careful layouting) to reduce their effects on the final performance.

Both MDACs in the signal paths are switched-capacitor circuits employing "flip-around" charge redistribution which benefits from larger feedback factors compared to a "non-flip-around" architecture [14]. They implement the algorithm expressed in Eq. 4.1 and its voltage transfer curve shown in Fig. 4.5.

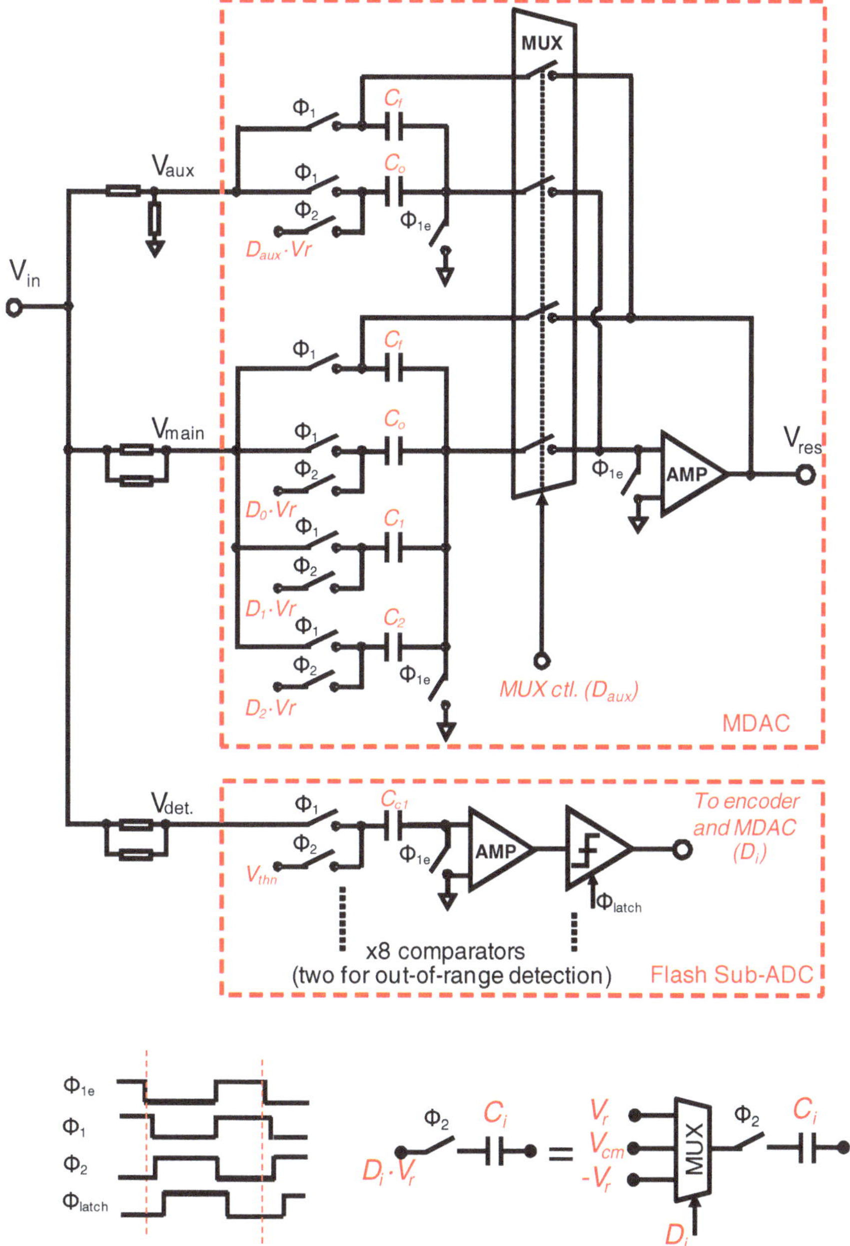

Fig. 4.4 Simplified schematic and timing diagram of the first pipelined stage with the parallel sampling technique (only single-ended version is shown for clarity)

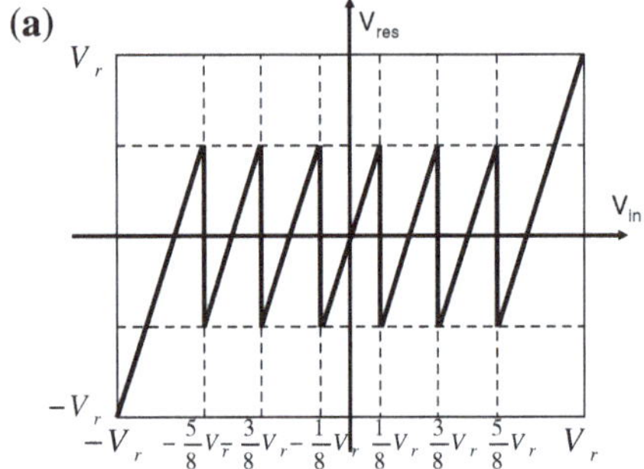

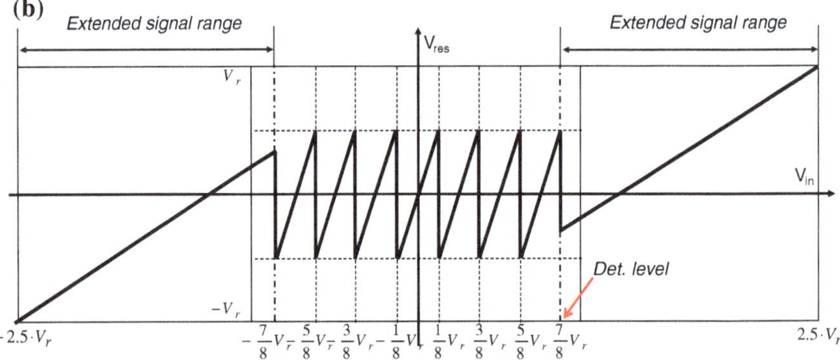

Fig. 4.5 First stage residue amplifier transfer curve: **a** conventional 2.5 b stage; **b** proposed stage with enlarged input range

$$\begin{cases} V_{res} = \frac{C_f+C_0+C_1+C_2}{C_f} \cdot V_{in} - \frac{(D_0 \cdot C_0 + D_1 \cdot C_1 + D_2 \cdot C_2)}{C_f} \cdot V_r & \text{if } -\frac{7}{8}V_r < V_{in} < \frac{7}{8}V_r \\ V_{res} = \frac{1}{A} \cdot \frac{C_f+C_0}{C_f} \cdot V_{in} - \frac{D_{aux} \cdot C_0}{C_f} \cdot V_r & \text{if } v_{in} < -\frac{7}{8}V_r \text{ or } V_{in} > \frac{7}{8}V_r \end{cases} \tag{4.1}$$

where V_{in} and V_{res} are the input and output signal of the stage respectively, and V_r is the reference voltage which is equal to 0.4 V. The MDAC input D_n corresponds to the output of the sub-ADC and it is controlled by the encoder of the Flash ADC as shown in Fig. 4.4. The total sampling capacitor size ($C_f + C_0 + C_1 + C_2$ in the main path and $C_f + C_0$ in the aux path) is chosen to meet the desired input referred thermal noise power requirement with respect to the full-scale voltage of $2V_{ppd}$. This ‘flip-around’ scheme achieves a closed loop gain of 4 with a feedback factor of ¼ in the main path and 2 and ½ in the auxiliary path respectively. In this stage, bottom-plate sampling techniques are used to reduce signal dependent charge injection of sampling switches and all the sampling switches are bootstrapped to improve linearity [8].

The residue amplifier uses a single folded cascode stage with gain boosting configuration which is similar to the one presented in [15, 16]. The schematic of the residue amplifier is shown in Fig. 4.6, where A_p and A_n are the gain-boosting amplifiers for the nMOS and pMOS cascode transistors respectively (schematics shown on the right of the figure). The CMFB block is a switched-capacitor common-mode feedback circuit [17], and Vbi are bias voltages at various nodes. The simulated DC gain of the amplifier (AMP) in the MDAC is about 70 dB with a gain-bandwidth higher than 1 GHz. The amplifier is designed to have an output swing of $0.8V_{ppd}$ using a single 1.2 V power supply.

The flash-ADC shown in Fig. 4.4 consists of eight comparators (each comprising a pre-amp and a regenerative latch) and a resistive reference ladder. Compared to that of a conventional 2.5 b stage, two additional comparators are needed to identify if the input signal is smaller or larger than the allowable input range of the main channel and to decide which channel should be connected to the residue amplifier. The decisions of the stage, for the various input ranges, are listed in Table 4.1.

As illustrated in Fig. 4.4, the MDAC is controlled by two-phase non-overlapping clocks which are denoted as Φ_1 and Φ_2, the sampling and the amplification phase, respectively. During Φ_1, the signal is tracked by sampling capacitors in both signal paths and the sampling network of the flash-ADC. The sampling actions in the main, auxiliary and detection paths are controlled by the same clock signal, and take place at the falling edge of Φ_{1e}; both V_{aux} and V_{main} are sampled onto the sampling capacitors simultaneously. Then, at the rising edge of Φ_{latch}, the sub-ADC decodes the signal level. After the decision is made, proper reference voltages ($\pm V_r$ or V_{cm}) are chosen and connected to the sampling nodes of the capacitors for subtraction.

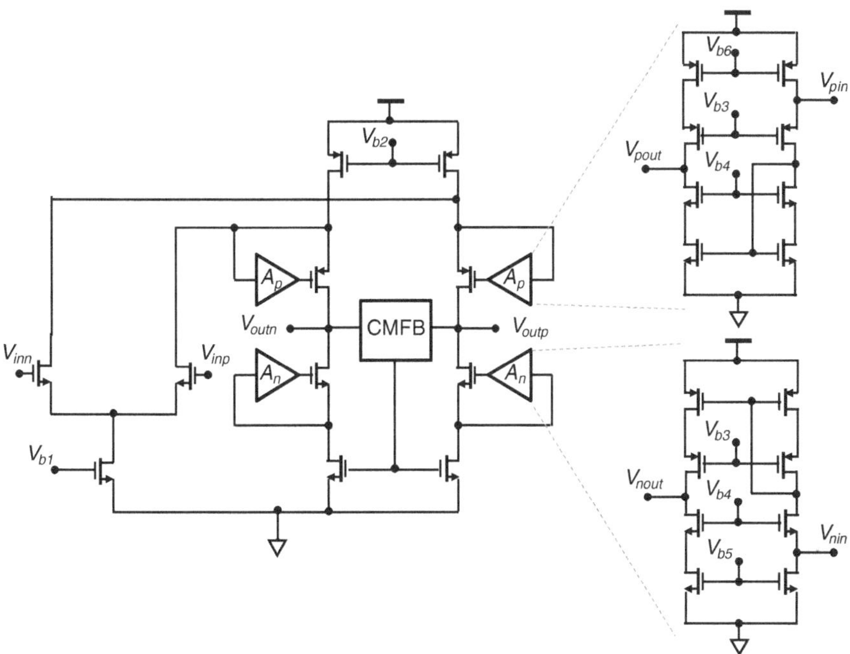

Fig. 4.6 Schematic of the residue amplifier in the first pipeline stage

Table 4.1 Decision of the first pipeline ADC stage

Analog input	Stage decision			
	D_0	D_1	D_2	D_{aux}
$V_{in} \geq 7/8V_r$				1
$5/8V_r \leq V_{in} \leq 7/8V_r$	1	1	1	
$3/8V_r \leq V_{in} \leq 5/8V_r$	1	1	0	
$1/8V_r \leq V_{in} \leq 3/8V_r$	1	0	0	
$-1/8V_r \leq V_{in} \leq 1/8V_r$	0	0	0	
$-3/8Vr \leq V_{in} \leq -1/8V_r$	−1	0	0	
$-5/8V_r \leq V_{in} \leq 3/8V_r$	−1	−1	0	
$-7/8V_r \leq V_{in} \leq -5/8V_r$	−1	−1	−1	
$V_{in} \leq -7/8V_r$				−1

At the same time, the feedback capacitor C_f of the main or auxiliary sampling network is selected and connected across the amplifier through the MUX as a feedback capacitor for charge redistribution and produces a residue signal for the following stages.

4.1.4 Simulation and Comparison

The performance of the proposed pipeline ADC was verified by behavioral simulations in Matlab. Linearity of the parallel-sampling first stage was studied by transistor level simulations in Cadence using TSMC 65 nm CMOS technology.

Only noise and clipping distortion were considered in the MATLAB simulations. A multi-carrier test signal was used to study the performance of the proposed ADC. As shown in Fig. 4.7a, the multi-carrier test signal has a signal amplitude distribution approaching a Gaussian distribution. The signal ranges covered by each signal path are also indicated: $0.8V_{ppd}$ for the main signal path and $2V_{ppd}$ for the overall ADC. The simulated SNCDR of the output signal of the main path, the auxiliary path, and the reconstructed signal with respect to input signal power are shown in Fig. 4.7b. The maximum SNCDR of both the main and auxiliary path is 59.9 dB. As observed in the figure, the SNCDR of the reconstructed signal follows that of the output signal of the main path when the input signal power is low, as most of the samples are processed by the main path. With the increase of input signal power, large amplitudes that are clipped in the main path are replaced by their attenuated versions from the auxiliary path, but the majority of the samples are still processed by the main path, hence the SNCDR of the reconstructed signal keeps increasing until the signal in the auxiliary path starts to clip excessively. The peak SNCDR of the reconstructed signal is 65.9 dB and about 92.5 % of the input signal amplitudes are processed by the main path. An improvement of about 6 dB in SNCDR and about 8 dB in dynamic range compared to its main path are observed. The SNCDR of the proposed ADC for a single sinusoid input signal is shown in

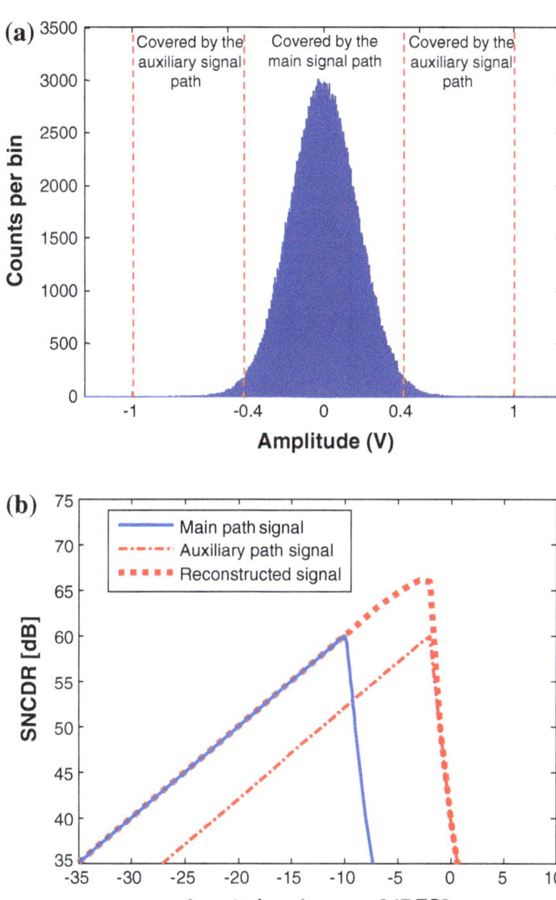

Fig. 4.7 **a** Histogram of a multi-carrier test signal; **b** SNCDR of the proposed pipeline ADC first stage for the multi-carrier test signal. (0 dBFS refers to the full scale signal power of the main signal path measured by a single sinusoid.)

Fig. 4.8. The parallel-sampling architecture does not improve the SNCDR for the single sinusoid signal due to its U-shaped amplitude probability distribution as discussed in Chap. 3, while it still has the advantage of improving the ADC dynamic range (by about 8 dB) which can be useful in some applications.

Transistor level simulations were carried out using a testbench built in Cadence analog design environment (ADE) for the purpose of verifying the linearity of the proposed pipeline ADC first stage. The schematics of this stage are shown in Fig. 4.4. Circuits of this stage were implemented using 65 nm CMOS technology. The stage operates at 200 MS/s and with a 1.2 V supply. A multi-tone test signal (consisting of 58 sinusoids with different phases with $PAPR > 10$ dB) was generated which possessed an approximately uniform spectrum over the bandwidth of interest, except for a narrow band of frequencies intentionally "missing", as shown in Fig. 4.9a. The output waveform spectra were analyzed to determine how much power had leaked into the "missing" band. The NPR is then calculated using

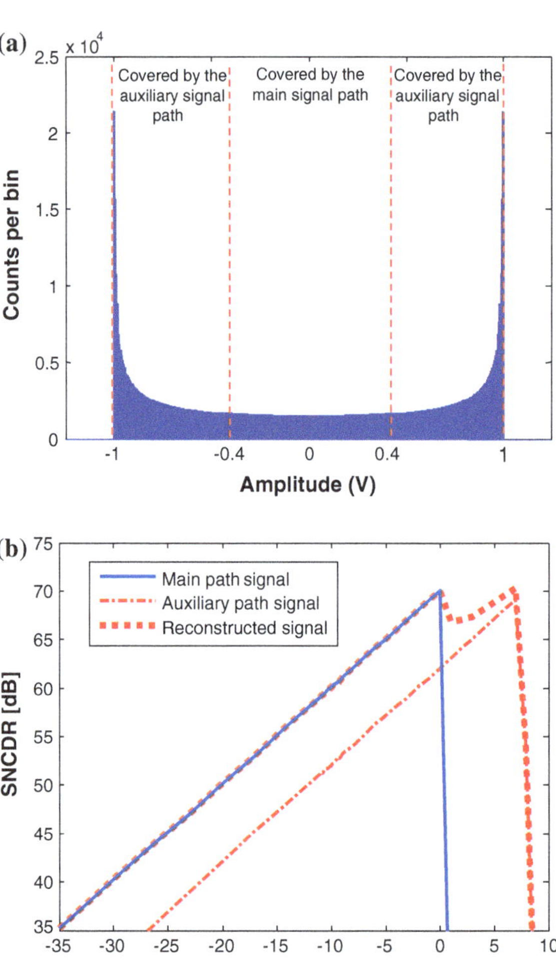

Fig. 4.8 **a** Histogram of a full-scale single sinusoid test signal; **b** SNCDR of the proposed pipeline ADC first stage for the single sinusoid signal

Eq. 2.1 in Chap. 2 (the signal power is measured at around 40 MHz in a bandwidth equals the notched frequency band at around 65 MHz). The output spectra (128p FFT with coherent sampling) of both signal paths in the first stage are shown in Fig. 4.9b, c. The simulation was done with only one of them enabled. Observed from the spectrum plots, the signal in the main path is heavily distorted because of its large amplitudes that exceed the linear signal range. This results in only 16 dB NPR (only distortion was taken into account). The auxiliary path processed an attenuated (by a factor of 2.5) version of the input signal which is within its linear signal range. Therefore, the NPR in the auxiliary path is much higher (about 75 dB). The reconstructed signal at the output of the first stage has an NPR of 67.9 dB, shown in Fig. 4.9d. Comparing it with the NPR of an ideal 12 b conventional ADC, which is 62.71 dB [18, 19], the simulated distortion power is below the noise floor of an ideal 12 b ADC by about 5 dB. Therefore, when the thermal noise power is

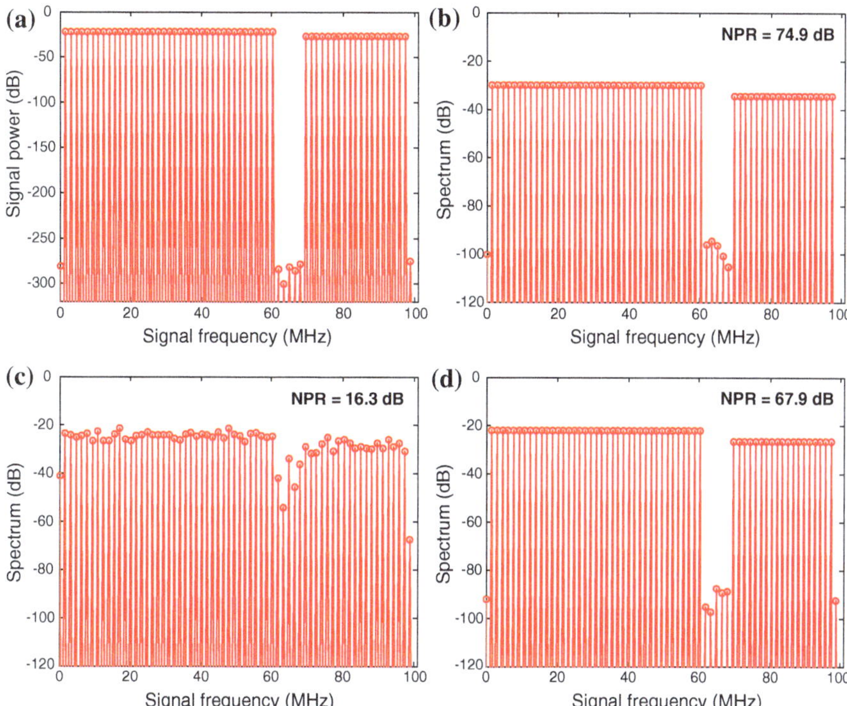

Fig. 4.9 Spectrum plots of **a** the test signal for transistor level simulation in Cadence, **b** the auxiliary path output signal, **c** the main path output signal, and **d** the reconstructed signal of the pipeline first stage. (Note: the signal power is measured at around 40 MHz in a bandwidth equals the notched frequency band.)

included (assuming SNR is limited by thermal noise), the total noise and distortion power in the "missing" band will be dominated mainly by the thermal noise.

As observed from simulations, the parallel-sampling first stage allows a 2.5 times increase of the input signal range without introducing excessive clipping distortion for the multi-carrier signal compared to the conventional pipeline stage shown in Fig. 4.1b. As the thermal noise power of the ADC is not signal dependent, the increase in signal power (thanks to the increase in signal range) improves the SNR of a thermal noise limited ADC proportionally. In this stage, the improvement in SNR is achieved without increasing the size of sampling capacitors, hence the power needed for the residue amplifier to drive the loading capacitors also stays the same and the additional circuits (an additional passive T/H, a MUX and two more comparators in the sub-ADC for out of range detection) consume mostly dynamic power. Therefore, the 6 dB improvement in SNR is achieved with less than half the power and area compared with the conventional approach of using larger devices to lower the thermal noise power (e.g. reducing the thermal noise power by 6 dB corresponds to 4 times increase in device sizes as well as a similar increase in power to maintain the same operation speed). For multi-carrier systems such as LTE-advanced Wi-Fi (IEEE802.11ac) [2, 3], equal system performance (in terms of SNR) can be achieved with the proposed ADC instead of a conventional pipeline ADC with one extra ENOB (considering thermal-noise limited scenarios).

4.2 Parallel-Sampling Architecture Applied to a TI SAR ADC

This section describes the architecture and operation of a 4 GS/s 11 b time-interleaving (TI) SAR ADC with a parallel-sampling frontend stage, which targets wideband direct sampling receivers for DOCSIS 3.0 cable modems [20]. The circuit design and simulations of the parallel-sampling frontend stage are presented.

4.2.1 A Hierarchical TI-SAR ADC Architecture

ADCs with GHz sampling rate and medium-to-high resolution (SNR > 50 dB) are mostly based on time-interleaving architecture nowadays [21–36]. In these ADCs, small sampling capacitors and transistors with low parasitic capacitance are desired in order to achieve high signal bandwidth and low power consumption. Therefore, these ADCs are normally thermal noise limited. Circuit techniques for maximizing signal swing are commonly adopted for the purpose of minimizing sampling capacitor sizes [21, 22, 25, 26, 29, 30, 32, 33].

Figure 4.10 shows a TI SAR ADC architecture (with hierarchical T/Hs) which is a suitable architecture for designing GHz sampling rate and medium-to-high resolution ADCs [21, 22]. The ADC presented in [21, 22] was designed for direct sampling of broadband multi-carrier signals in cable applications. It consists of four frontend T/Hs (including also interface circuits comprising demultiplexer and redistribution buffers) and four quarter ADCs (QADC) each consisting of 16 SAR ADCs (sub-ADC units). The general operation is described as follows: the input signal is sampled first by four frontend T/Hs in a time-interleaved fashion, and then the output signal of each T/H is buffered and resampled by T/Hs of the sub-ADC and quantized by the sub-ADC, again in a time-interleaved fashion. Finally, the outputs of the sub-ADCs are combined to make one high speed digital output. The ADC employing this architecture has demonstrated a high sampling rate of 3.6 GHz and a thermal-noise-limited SNR of 54 dB up to 1 GHz [22]. In this ADC architecture, the speed requirement of its sub-ADCs is relaxed by introducing parallelism, but the challenge is shifted to the ADC's frontend stage (T/Hs, demultiplexers and redistribution buffers) which have to sample the high frequency input signal and redistribute the sampled signal to each sub-ADC with enough accuracy at GHz sample rate. In [21, 22], multiple techniques were used in the frontend stage in

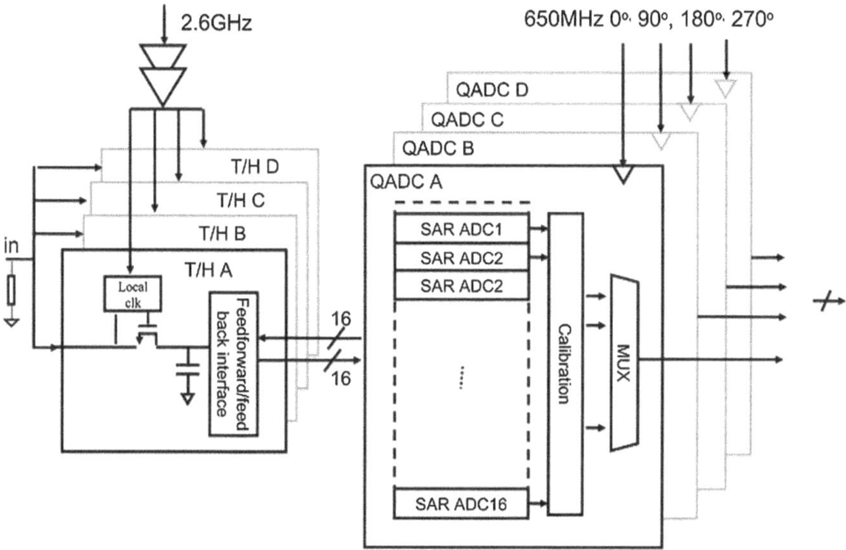

Fig. 4.10 A hierarchical TI-SAR ADC architecture [21]

order to achieve the target performance. Besides the hierarchical sampling and demultiplexing which allows only one sub-ADC to be connected to the frontend T/H at a time to reduce the capacitive loading of the front-end T/Hs, an innovated approach named "feedforward-sampling feedback-SAR" is introduced to alleviate the linearity limitation of redistribution buffers interfacing between frontend T/Hs and sub-ADC units and enhance the linear input signal range of the ADC.

The parallel-sampling architecture presented in Chap. 3 can be applied to the frontend stage of this ADC to improve its signal range further with the purpose of improving the SNR with better power efficiency for multi-carrier signals, as will be discussed in the following sections.

4.2.2 A Parallel-Sampling Frontend Stage for a TI-SAR ADC

Due to the stringent tradeoff between thermal noise, speed and power of an ADC as discussed in Chap. 2, improving the SNR of a thermal noise limited ADC by reducing the noise power requires increasing sampling capacitors and devices size exponentially. To lower the thermal noise power of the TI-ADC shown in Fig. 4.10 by 6 dB requires at least 4 times increase in power and area, as well as a significant increase in design complexity to deal with the increased interconnection capacitances (signal and clock distribution) to maintain a similar bandwidth. As the ADC is designed for a broadband receiver for a DOCSIS 3.0 cable modem [20], the received signal is a multi-carrier signal which can be composed by up to 126 256-QAM modulated sub-carriers. Therefore, the parallel-sampling technique can be applied to this ADC to reduce the increase in power consumption needed to achieve a better SNR.

Figure 4.11 shows the block diagram of TI-ADCs with and without a parallel-sampling frontend stage. The ADC shown in Fig. 4.11a has the same architecture as the one shown in Fig. 4.10, but only one T/H and one QADC are shown for clarity. To apply the parallel-sampling architecture to this ADC, an additional signal path (auxiliary signal path) and a range detection path are introduced as shown in Fig. 4.11b. The additional circuits include a signal scaling block and a frontend T/H, a range detection block for each QADC, as well as an additional interface buffer, backend T/H and path selection MUX (SelMUX) for each sub-ADC unit. The comparator, DAC and SAR controller in each sub-ADC unit are shared by the main and auxiliary signal paths thanks to the adaptive signal paths selection.

In the ADC shown in Fig. 4.11b, the frontend T/H of the main signal path samples a two times larger input signal compared to that of its auxiliary signal path

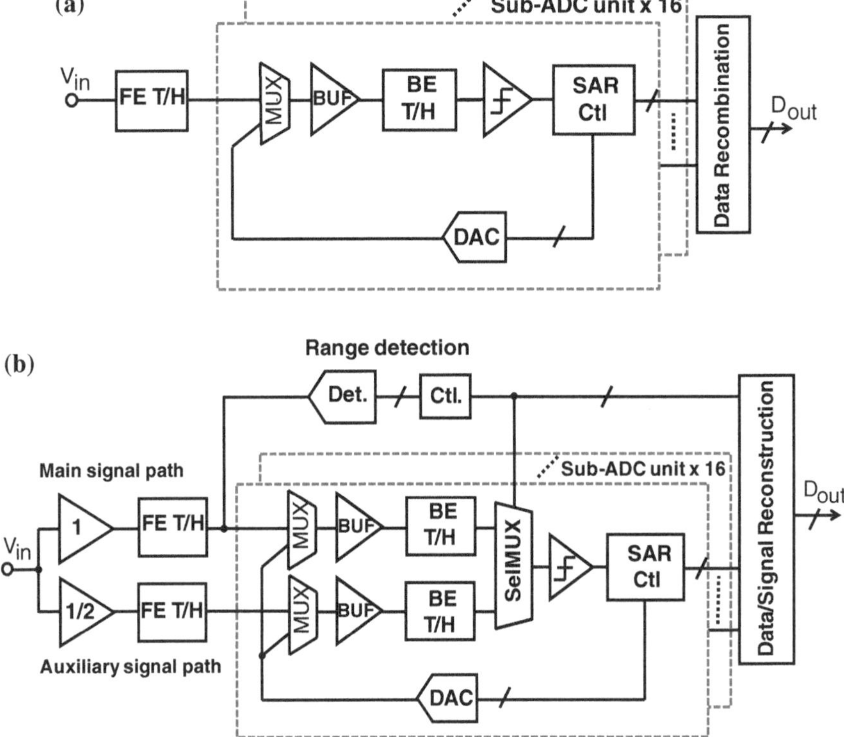

Fig. 4.11 **a** The ADC architecture without a parallel-sampling frontend stage; **b** the proposed parallel-sampling ADC architecture. (Only one FE front-end T/H and one QADC are shown for clarity.)

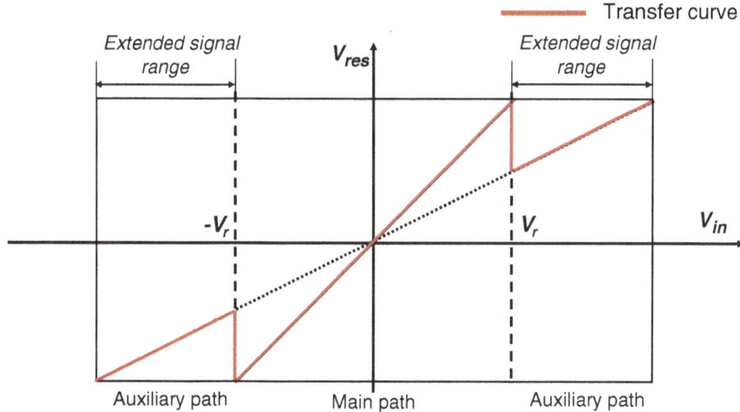

Fig. 4.12 Signal transfer curves of the proposed frontend stage

and allows clipping large signal amplitudes; while the T/Hs of the additional signal (auxiliary) path sample an attenuated version of the input signal. Signals in both signal paths are sampled simultaneously. When the sampled signal in the main path exceeds the allowed signal range which is detected on a sample-by-sample basis, the sampled signal in the auxiliary signal path is selected for further conversion by the sub-ADC. The input and output transfer curve of the proposed parallel-sampling stage are shown in Fig. 4.12, where V_{in} and V_{res} are the input and output signal of the stage respectively, and $[-V_r, V_r]$ is the input range of the main signal path. With an attenuation factor of two in the range scaling block for the auxiliary signal path, the parallel-sampling architecture enlarges the input range of the ADC by a factor of two compared to the previous design shown in Fig. 4.11a. The effective input range of the ADC is larger than the linear range of its frontend T/Hs and sub-ADC units.

Implementation and simulation of the proposed parallel-sampling frontend stage are presented in the next section. The ADC is designed using CMOS 40 nm technology and operates with dual supply voltages of 1.2 V and 2.5 V. The goal of the design is to improve the STNR by 6 dB compared to the previous work reported in [22] and to achieve an aggregate sample rate of 4 GHz.

4.2.3 Implementation and Operation

Figure 4.13 shows a schematic representation of the parallel-sampling frontend stage of the proposed ADC. The actual circuit implementation is fully differential. On the left of the schematic is the signal range scaling stage which was implemented by polysilicon resistors. This stage provides an attenuated version of the input signal (by a factor of 2) for the auxiliary path by a resistive divider and an unattenuated version for the main signal path with dummy resistors in the signal

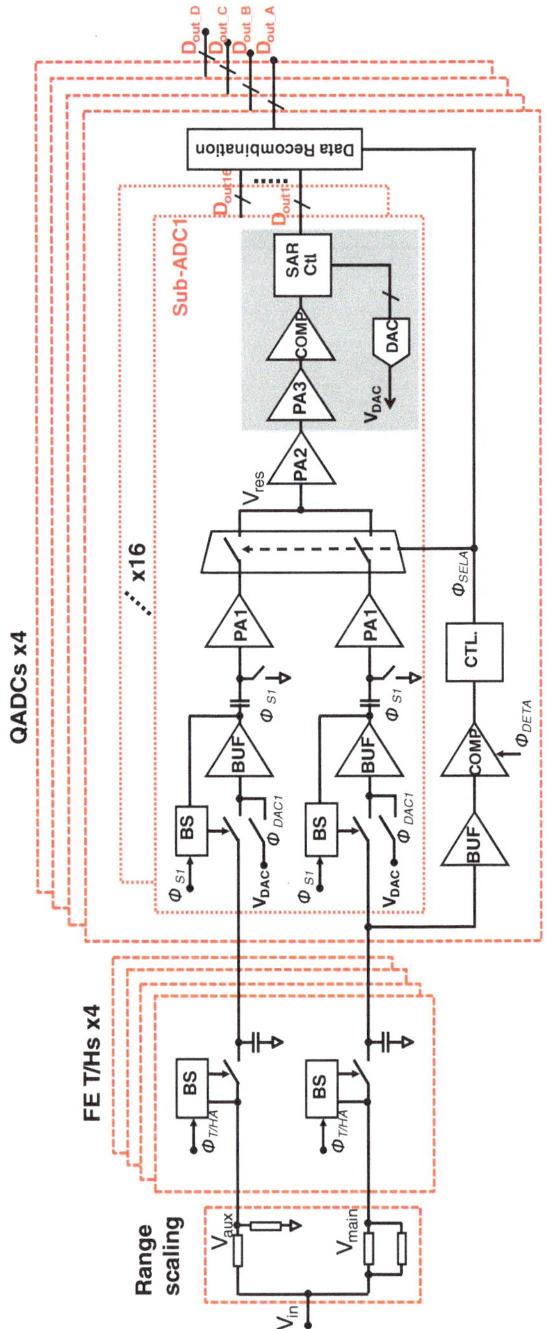

Fig. 4.13 Schematic diagram of the parallel sampling frontend stage for the TI-SAR ADC

path for the purpose of matching the bandwidths of the two signal paths. This stage also provides a 50 Ω input termination. The unit resistors are sized to have an intrinsic matching of 12 b accuracy. The range scaling stage is followed by the frontend T/Hs, demultiplexers, source follower buffers and backend T/Hs in each sub-ADC unit. These circuit blocks are the same as those described in [21, 22]. To ensure high linearity at GHz sample rate, the switches in both signal paths are bootstrapped to reduce the signal dependent modulation of their on-resistance and to enhance their bandwidth for large signal amplitudes. The backend T/H is then followed by preamplifiers which are implemented by differential pairs with reset function to reduce memory effects between samples. The reset is controlled by a semi-synchronized (rising edge asynchronized while falling edge synchronized) logic circuit similar to the one reported in [37] which senses the output of the comparator and activates the reset switch as soon as the comparator makes a decision. The MUX for the adaptive signal paths selection is placed between the first preamplifier and the second preamplifer to avoid disturbing the high speed sampling operation which would lead to linearity degradation of the sampled signal (such as reducing tracking time and unwanted charge sharing). Schematics of the preamplifiers and MUX are shown in Fig. 4.15. The range detection block is implemented as a two-level flash-ADC consisting of two comparators (each comprising a pre-amp and a regenerative latch) and is similar to the one described in Sect. 4.1 of this chapter (Fig. 4.4).

The linear input signal range of the T/Hs and the sub-ADC (defined by its DAC) is designed to be $1.2V_{ppd}$. The proposed parallel-sampling frontend stage enables an effective input signal range of $2.4V_{ppd}$. Sizes of sampling capacitors and transistors are chosen to achieve an STNR of 55 dB with a $1.2V_{ppd}$ full scale sinusoid input signal. With an enlarged input signal range of $2.4V_{ppd}$, the input signal power is boosted by 6 dB. Therefore, the STNR of the proposed ADC will also improve by about 6 dB for multi-carrier signals.

Figure 4.14 shows the timing diagram (only that of the frontend T/H and one sub-ADC shown for clarity). The operation of the parallel sampling stage is explained as follows. The frontend T/Hs in both signal paths are sampled synchronously at 1 GSps. With 4 times interleaving, the total sample rate is 4 GHz.

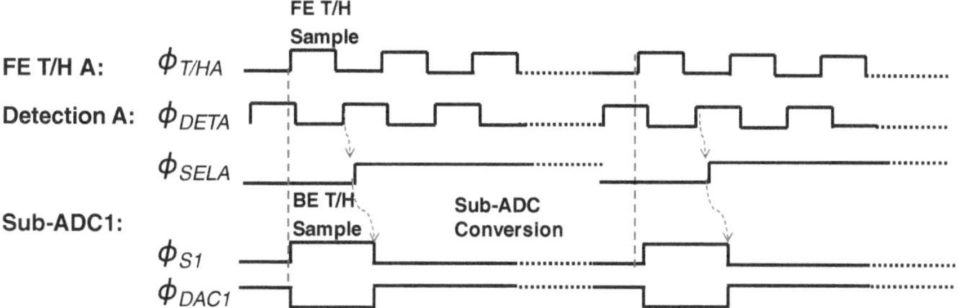

Fig. 4.14 The timing diagram of the proposed parallel-sampling frontend stage. (Only shows that of the front-end T/H A and sub-ADC1 in QADC A for clarity.)

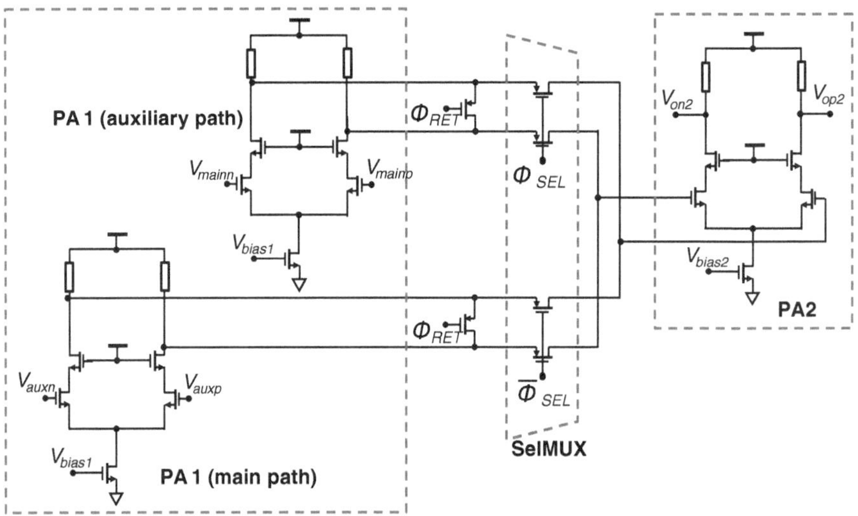

Fig. 4.15 Schematic of the first two preamplifers and the MUX

When $\Phi_{T/HA}$ is high, the input signal is tracked by the sampling capacitors of the frontend T/Hs in both signal paths. The sampling actions are taking place at the falling edge of $\Phi_{T/HA}$; both V_{aux} and V_{main} are sampled onto the sampling capacitors simultaneously. Then, at the rising edge of Φ_{DETA}, the sub-ADC detects the signal level. In the same clock period, the sampled signals of the frontend T/Hs are also resampled to the backend T/Hs of one of the sub-ADC units (when Φ_{S1} is high). After the comparators in the detection path make a decision, Φ_{SELA} is activated and the proper signal path is connected to the input of the second preamplifier (PA2). At the falling edge of $\Phi_{S1,}$ the resampling phase is complete. Then, by Φ_{DAC}, the DAC output is connected to the input of the buffer to start the SAR conversion.

4.2.4 Simulation Results

A testbench was built in Cadence analog design environment (ADE) to verify the linearity performance of the proposed parallel-sampling frontend stage for the GHz sample rate TI-SAR ADC. A multi-tone signal (comprising 55 sinusoids with different phases and a 'missing' band) occupying a bandwidth up to 1.9 GHz was generated as a test signal. The test signal has a PAPR of 11.4 dB. As mentioned earlier, this stage was designed to have an input signal range of $2.4V_{ppd}$ enabled by the parallel-sampling architecture (the linear input signal ranges of frontend T/Hs and sub-ADCs were designed to be $1.2V_{ppd}$), and each time-interleaved T/Hs samples at 1 GS/s resulting in an aggregate sampling rate of the stage of 4 GHz. As the purpose was to study the linearity performance of the frontend circuits (including the range scaling block, frontend T/Hs, redistribution buffers and backend T/Hs in the sub-ADCs), some circuit blocks in the sub-ADCs and clocking paths (DACs, comparators, SAR controllers, signal reconstruction block and multi-phase clock generation blocks) were implemented with ideal components (verilog-A models) to speed up the simulation time without affecting the main investigation. The NPR was used to characterize the linearity of the proposed parallel-sampling frontend stage for broadband multi-carrier signals.

Figure 4.16a shows the time domain waveform of the test signal and (b) shows the corresponding output of the detection block which indicates whether the sampled values are out of the desired range or not. In Fig. 4.17, the spectra of the sampled signals in both signal paths of the frontend stage are shown (coherent sampling and 128p FFT), as measured at the outputs of the frontend T/Hs. The simulation was done with one of signal path enabled each time and only distortion was taken into account. As can be observed from the spectrum plots, the NPR of the signal processed by the main path is only 16.7 dB due to excessive clipping distortion; while the NPR in the auxiliary path is about 58.2 dB, which is much higher thanks to the attenuated signal it processed. The average distortion powers measured in the "missing" band of the signal spectra in the auxiliary and main paths are −84.5 and −37 dBFS respectively. When both the signal paths are enabled, the output signal after reconstruction has an NPR of about 57.9 dB and the average distortion power measured in the 'missing' band is −78.5 dBFS, as shown in Fig. 4.17. The thermal noise power of the main signal path was designed corresponding to that of an 10-b ideal ADC which has an ideal NPR of 51.6 dB for an

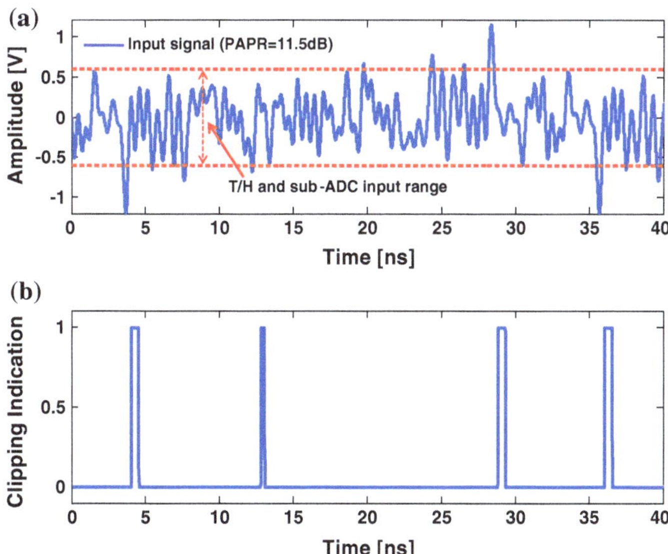

Fig. 4.16 **a** Time domain waveform of the test signal and **b** the output of the detection path (indicate 'high' when clipping is detected)

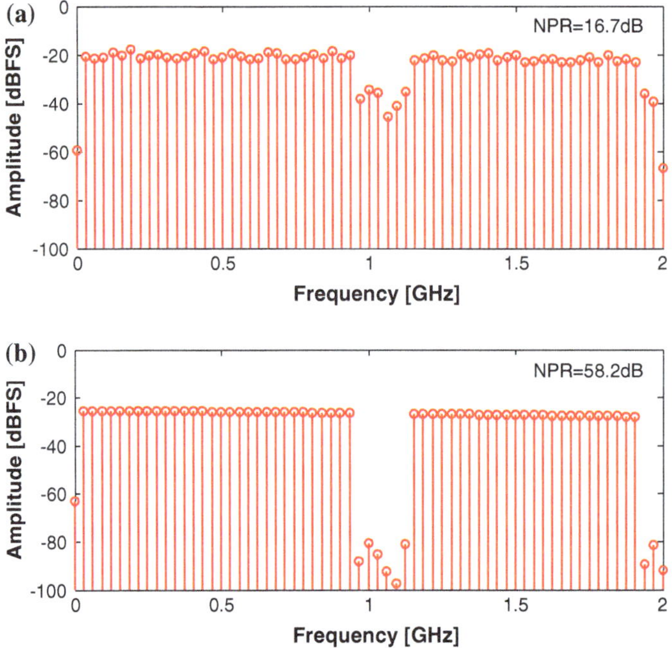

Fig. 4.17 Spectra of sampled signals of **a** the main path and **b** the auxiliary path. (The simulation was done with only one of them enabled each time.)

input signal having Gaussian amplitude distribution [18, 19]. Comparing the simulated NPR of the reconstructed signal with that of an ideal 10-b ADC, the simulated distortion power in the "missing" band is below the designed thermal noise floor by about 6 dB. Therefore, the total noise and distortion power in the "missing" band will be dominated mainly by the thermal noise. Comparing Fig. 4.18 with Fig. 4.17, we can observe that the signal power of the reconstructed signal is similar to the signal power at the output of the main path, but the simulated NPR is about 38 dB better.

The simulation results confirm that this parallel-sampling frontend stage is able to process a broadband multicarrier signal with a PAPR of 11.4 dB and $2.4V_{ppd}$ amplitude without causing severe distortion (although the frontend T/Hs and sub-ADCs are designed to have only a 1.2 V linear input signal range). As the input signal power is boosted by 6 dB while the distortion power is kept below the thermal noise floor by 6 dB, both STNR and SNDR are hence improved by about 6 dB (excluding jitter noise power).

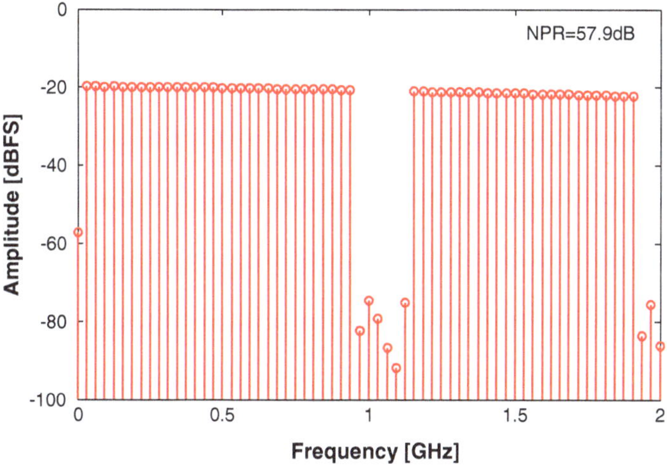

Fig. 4.18 Spectrum of the reconstructed signal

4.3 Design of a 1 GS/s 11-b Parallel-Sampling ADC for Broadband Multi-Carrier Systems

As a first step toward building a 4 GS/s 11-b parallel-sampling ADC for a wideband direct sampling receiver used in a DOCSIS 3.0 cable modem as presented in Sect. 4.3, a prototype IC of a 1 GS/s 11-b parallel-sampling ADC was implemented, which can be time-interleaved by four times to achieve an aggregate sample rate of 4 GS/s in future work. This test-chip was implemented in 65 nm LP CMOS. Many circuit building blocks (sub-ADC, clock generation, biasing, digital calibration circuits, etc.) in this test-chip were reused from previous works [21, 22] for the purpose of a fast proof of the parallel-sampling ADC concept. Besides for proof-of-concept, this test-chip was also designed for the purpose of studying the performance of the range scaling circuit, high-speed frontend T/H circuits and the run-time out-of-range detection circuit operating at GHz sample rate. To reduce the design complexity, the adaptive signal paths selection and the shared backend ADC were not implemented in this test chip. Instead, two 1 GS/s time-interleaving ADCs are used separately to convert the main and auxiliary signals and the outputs of the ADCs are combined off-chip. In this section, the implementation and experimental results of this prototype IC are presented. The architecture of the proposed ADC IC is discussed in Sect. 4.3.1; a description of the circuit implementations is given in Sect. 4.3.2; layout and test-chip implementation are shown in Sect. 4.3.3; measurement setup and experimental results of the prototype ADC are addressed in Sects. 4.3.4 and 4.3.5; performance summary of the prototype IC and comparison with other published works are given in Sects. 4.3.6 and 4.3.7.

4.3.1 Architecture and Operation Overview

The architecture of the prototype IC is shown in Fig. 4.19. It consists of two sub-ADCs (main and auxiliary) and a range detector in parallel. The sub-ADCs are preceded by a range-scaling stage and their outputs are combined digitally. Each of the sub-ADCs contains a dedicated front-end T/H and 16 SAR-ADC units. The principle of operation of this proposed ADC is as follows. The range scaling stage splits the front-end analog input signal (V_{in}) into two signals which are scaled versions of each other and are sampled by two sub-ADCs simultaneously. The input signal range of the main ADC is 2 times larger than the linear input range of front-end T/H and the output range of DAC in the sub-ADC (within SAR-ADC unit). The signal processed by the auxiliary ADC is an attenuated version of V_{in} (by a factor of 2). Large amplitudes that clip the main ADC are detected by the range detector on a sample-by-sample basis, and during signal reconstruction, the clipped samples from the main ADC are replaced by corresponding samples from the auxiliary ADC after they are amplified digitally. This allows the ADC to handle a significantly larger input signal range without severe clipping distortion while offering a better overall SNDR than its sub-ADCs as explained in Chap. 3.

4.3.2 Circuit Implementation

The ADC is based on a time-interleaved ADC architecture with hierarchical T/Hs [21, 22], as shown in Fig. 4.19. The main building blocks of this ADC are an input

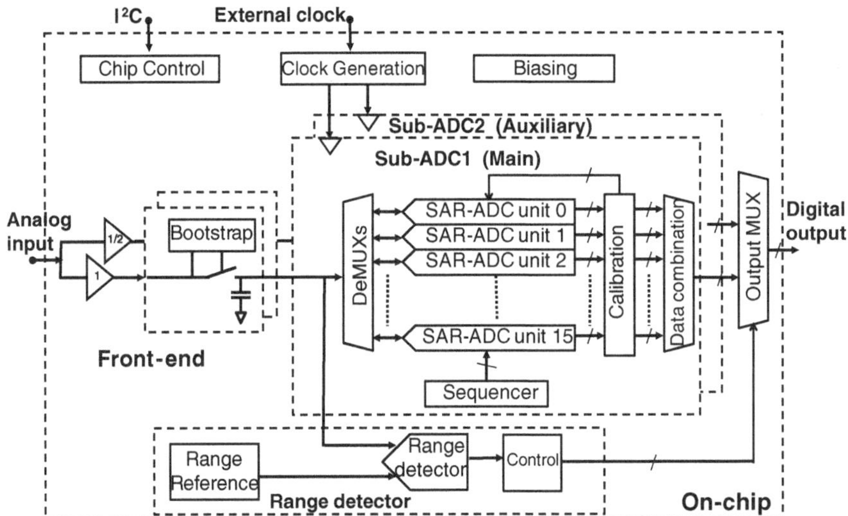

Fig. 4.19 Block diagram of the prototype IC

interface, a range-scaling stage, a hierarchical sampling network, two sub-ADCs each consisting of 16 time-interleaved SAR-ADC units, a range detector, digital circuits for on-chip calibration, and additional circuits required to interface the prototype IC with test equipment including control interface, clock buffers and output drivers. This section presents the design of these circuit building blocks.

4.3.2.1 Input Interface and Range-Scaling Stage

The input interface provides an on-chip input termination (100 Ω differentially) and sets the input common-mode voltage of the ADC. Figure 4.20 shows two configurations which are suitable for high speed ADCs. The one on the left uses two resistor dividers in a single-ended configuration to provide input termination and common-mode voltage setting. Both differential and common-mode signal have the same impedance to ground (50 Ω). In this configuration, the ground node acts as another signal input in the layout design. If the two resistor dividers are not well matched and their ground node is noisy or carries other interferences, these signals can be coupled directly into the signal path resulting in SNR degradation. The circuit on the right employs two serial resistors placed across the inputs of the ADC to set the input termination, and the common-mode voltage is supplied from the center point of two identical resistors. As the common-mode voltage is generated by a common reference ladder or buffer, any noise and interference coupled in occurs as a common-mode signal and can be suppressed by the ADCs common-mode rejection. In this design, the configuration shown in Fig. 4.20b was selected due to the advantage mentioned above and a capacitor of about 10 pF was added to the common-mode node (V_{cm}) to provide a low impedance path to ground for minimizing the high frequency charge kick back due to the sampling switches operating at GS/s.

The range-scaling stage splits the front-end input signal (V_{in}) into two signals which are scaled versions of each other. In this stage, achieving an accurate scale ratio and at the same time bandwidth matching of two signal paths (main and

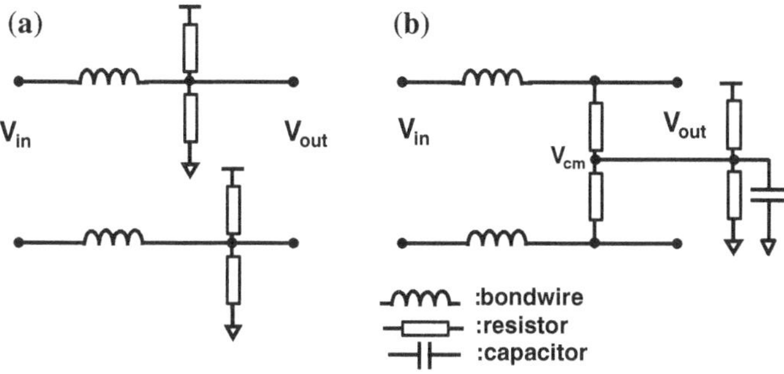

Fig. 4.20 Two circuit configurations for the input interface

auxiliary) are important for the performance of this parallel-sampling ADC operating at GHz frequencies. Figure 4.21 shows three circuit configurations that are considered for the range scaling stage. The circuit configurations shown in Fig. 4.21a are implemented by active components and in Fig. 4.21b by passive components. An active solution requires high gain amplifiers with feedback control to make an accurate gain. The requirements of such amplifiers are similar to that of the front-end sample-and-hold-amplifier (SHA) of a typical pipelined ADC or the input buffer of a high-speed and high-resolution ADC which can consume 30–40 % of the total power consumption of an ADC [9, 14]. It is challenging to achieve an accurate gain matching of two signal paths of better than 11-b and signal bandwidth beyond GHz with current technology. Besides that, such an amplifier will also pose extra limitations on the noise, linearity and dynamic range of the ADC. The passive solution shown in Fig. 4.21b uses a resistive divider connected between the differential inputs of the ADC to attenuate the input signal and doesn't consume extra power. The resistor ladder doesn't affect the noise performance of the ADC; the sampling noise stays the same (*kT/C*) but at a cost of reduction in signal bandwidth. An accurate scale ratio can be realized by intrinsic matching of poly-resistors without post trimming in current technology [15] and a large input bandwidth of multi-GHz can be achieved simultaneously.

In this design, the passive solution is selected in favor of low complexity and low power consumption. The range-scaling stage is implemented by a poly resistor divider having an attenuation factor of 2 for the auxiliary signal path, and it also contributes to part of the input termination as shown in Fig. 4.22. A few measures are taken in this design to address the gain and bandwidth matching issue of the two signal paths for achieving good performance at GHz sampling rate: firstly, the poly resistors divider is designed with an intrinsic matching of approximately 12-b to minimize the gain mismatch between the two signal paths; secondly, dummy

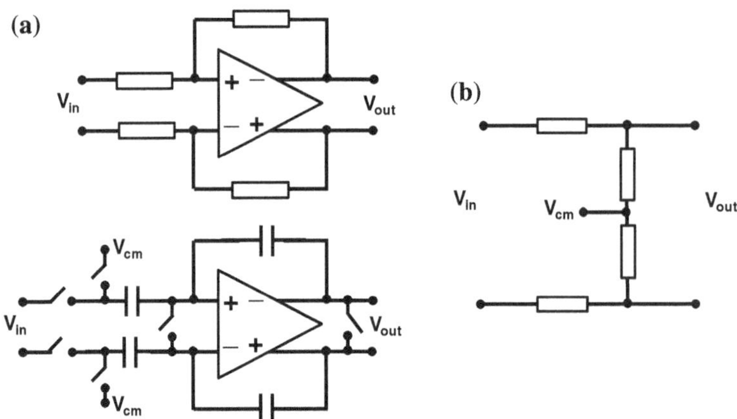

Fig. 4.21 Circuit configurations for the range-scaling stage; **a** with active components; **b** with passive components

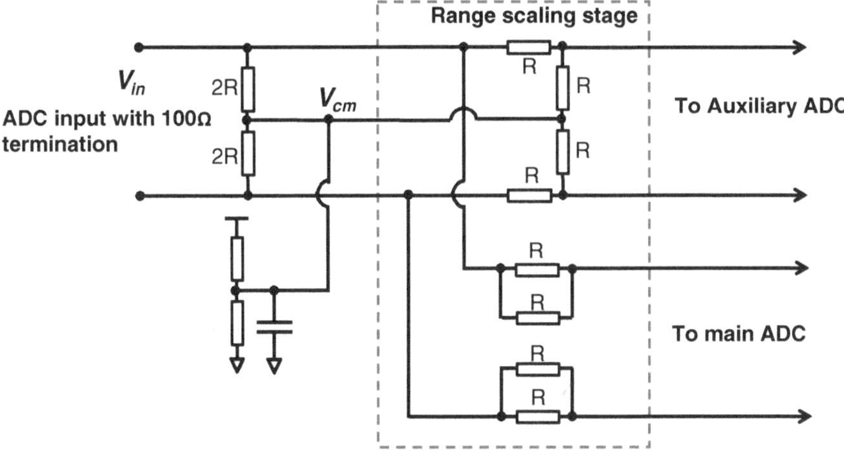

Fig. 4.22 Schematic of the input interface and the range-scaling stage

resistors are added in the main ADC input signal path to match the RC time constant with that of the auxiliary ADC input signal path; thirdly, the two input signal paths in the layout are designed as equal length traces with similar surrounding.

4.3.2.2 The Sub-ADC Architecture and the Sampling Network

With the purpose of achieving a GHz sampling rate and an SNDR better than 50 dB with good energy efficiency, a time-interleaved ADC architecture was adopted for main and auxiliary sub-ADCs, as shown in Fig. 4.23.

In a time-interleaved ADC, mismatches among its parallel ADC units, such as gain, offset, timing and bandwidth mismatches, degrade the overall SNDR [38–42]. Gain and offset mismatches are relatively easy to be minimized with power efficient calibration circuits, while minimizing bandwidth and timing mismatches for a TI-ADC with GHz sampling rate is very challenging because they are difficult to be detected and corrected [41, 42]. In this design, offset and gain mismatches among

Fig. 4.23 The sub-ADC architecture

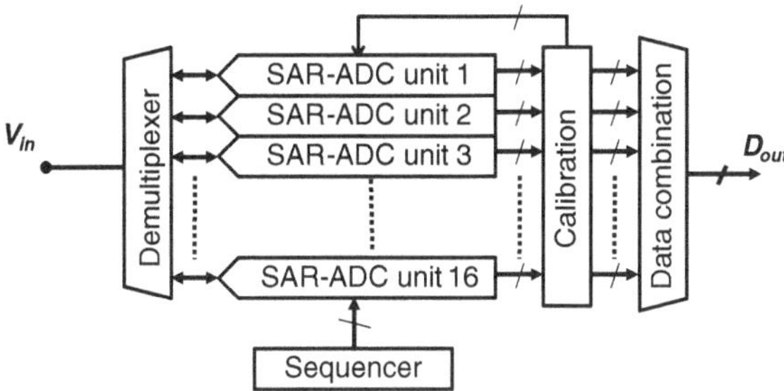

the interleaved ADC units are minimized through on-chip calibration circuits, while the bandwidth and timing mismatches are minimized by design and careful layout.

The bandwidth and timing matching of a time-interleaving ADC is dictated by its sampling network which makes it the most critical circuit block of a GHz sampling rate ADC. In this design, in order to achieve clock frequencies up to 1 GS/s while simultaneously maintaining an overall high linearity at high signal frequencies, a hierarchical sampling topology for the sampling network is chosen [21, 22]. Figure 4.24 shows the simplified schematic of the sampling network of the proposed ADC with parallel-sampling architecture. Most of the circuit building blocks are reused from the previously published works [21, 22] with improvements in the bootstrapping circuits and layout of the sampling capacitors for better linearity. Each of the main and auxiliary sampling networks consists of a front-end T/H and a 1-to-16 demultiplexer. The input signal is sampled by the front-end T/H and then re-sampled by the back-end T/Hs which are part of the SAR ADC units. The back-end T/Hs are operated in a time-interleaved fashion. Although the hierarchical sampling network comes with a power and noise penalty due to the need of buffering the sampled signal and redistributing it to the backend T/Hs, compared to a sampling network without hierarchy, the requirement of a high input signal bandwidth and high frequency linearity in this design makes it a desirable choice. Major benefits of the hierarchical sampling topology are as follows: firstly, since the timing is only decided by a common front-end T/H, the timing mismatch among the interleaved ADC units, which is a major problem of a time-interleaving ADC, is avoided; secondly, the physical separation of the front-end T/H from the back-end T/H by a buffer simplifies the design of the input signal and clock distribution network, and the signal and clock interconnect of the front-end T/H is reduced drastically, which enables a large input bandwidth and low clock jitter noise. Having a larger input bandwidth than the signal bandwidth allows reducing the impact of bandwidth mismatch on the SNDR of the ADC [38, 39]. In this design, the sampling capacitor of the front-end T/H is only 100fF which can be driven with good linearity at GHz sample rate without an on-chip buffer. In this design, the bandwidth of the front-end T/H is improved further by introducing a demultiplexer between the front-end T/H and back-end T/Hs. The demultiplexer is composed of an array of switches and is controlled in such a way that only one back-end T/H preceded by a source follower buffer loads the front-end T/H at any given moment in time [21, 22]. This allows minimizing the loading of the front-end T/H, hence the bandwidth and linearity are improved and many back-end T/Hs are allowed to be connected to a single front-end T/H.

Besides timing and bandwidth matching, the ability of linear handling of a large input signal swing at GHz sampling rate is also important for achieving a high SNDR in this design. A major issue is the impedance modulation of the sampling switches as a function of the signal amplitude. Therefore, the switches in the front-end T/Hs and the demultiplexer are bootstrapped to reduce their impact for handling large signals with good linearity and to reduce the signal dependent kickback charge to the ADC input. The schematic of the bootstrapping circuit for the

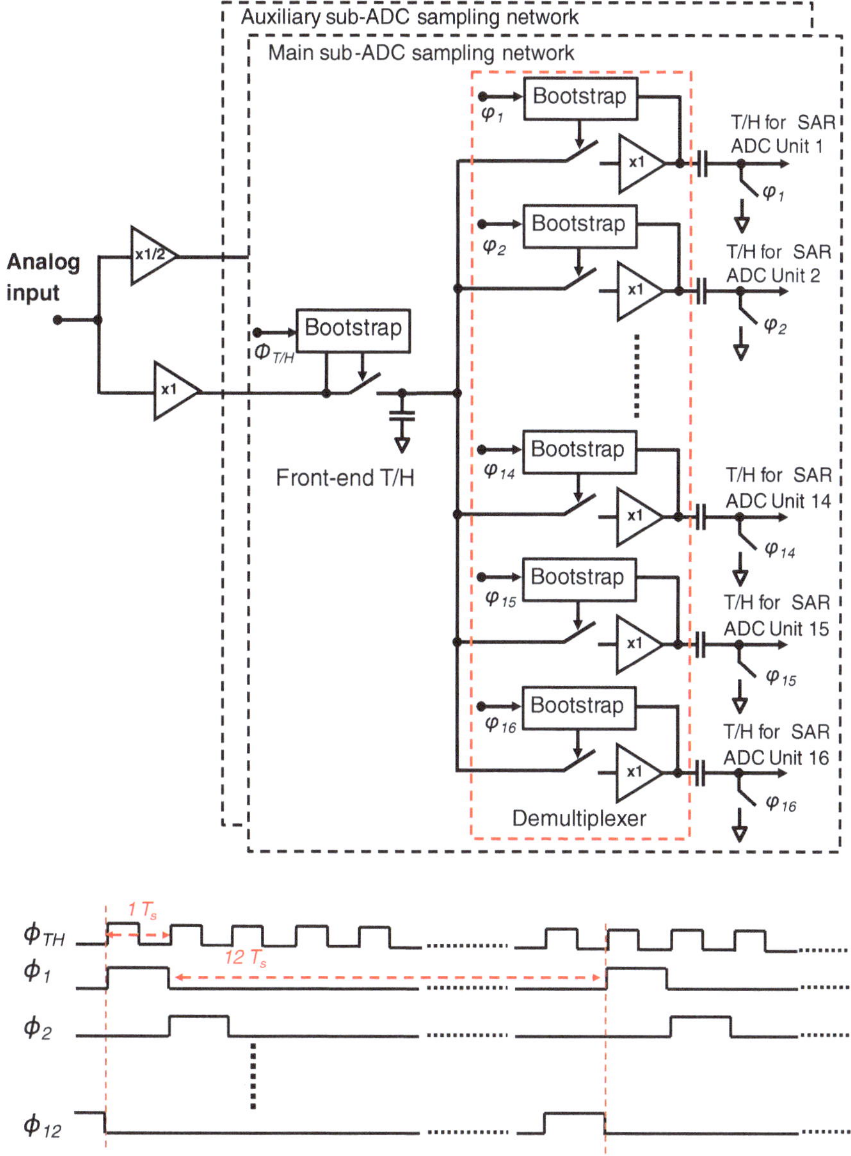

Fig. 4.24 Schematic diagram of the sampling network and its timing diagram

front-end T/Hs is shown in Fig. 4.25. It is a modified version of the bootstrapped switch proposed in [21].

The bootstrapping circuits for switches in the demultiplexer are similar with that of the front-end T/H, but their inputs are buffered by source follower buffers to avoid charge stealing from the sampling capacitor of the front-end T/H.

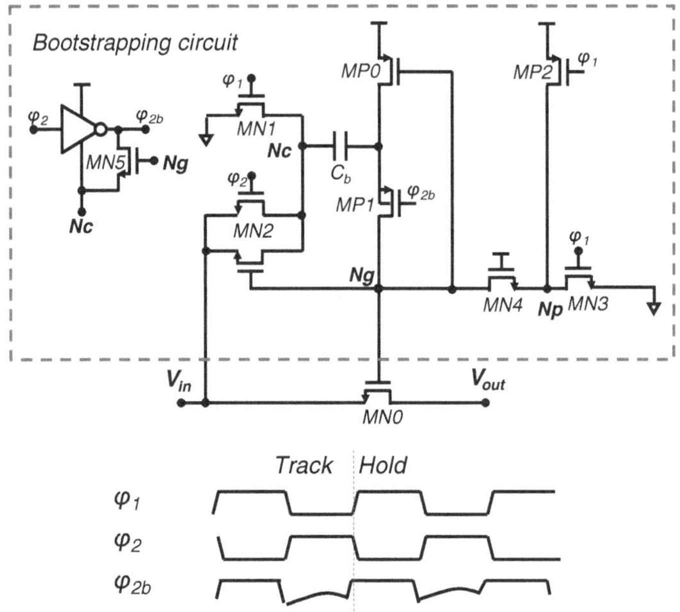

Fig. 4.25 Schematic of the bootstrapped track-and-hold circuit and its timing diagram

4.3.2.3 The SAR ADC Unit

In this prototype, the SAR ADC unit is reused from the previous works which has been validated by measurements [21, 22]. The architecture of the SAR ADC unit is shown in Fig. 4.26. It consists of a source follower buffer, a T/H, a comparator with three cascade differential pairs as its preamplifiers, a SAR digital controller, a current steering DAC (CS-DAC) and two calibration DACs for correcting gain and offset errors. The source follower buffer and the T/H of the SAR ADC unit are shared with the sampling network, as shown in Fig. 4.26.

The operation of the SAR ADC is based on a feedforward-sampling feedback-SAR principle [21, 22] which eliminates the distortion stemming from the buffer connecting the front-end T/H with the back-end T/Hs. During the sampling phase (when ϕ_{TH} is high), the T/H of the SAR ADC unit is connected to the front-end T/H through the demultiplexer and the source follower buffer, while the output of the DAC is disconnected from the input of the buffer. The signal sampled by the front-end T/H is then resampled to the sampling capacitor of the SAR-ADC in the resampling phase (when ϕ_{resamp} is high). During the SAR conversion phase (when ϕ_{DAC} is high), the front-end T/H is disconnected from the input of the source follower buffers; instead the output of the DAC is connected to the input of the buffer. The difference between the sampled signal and the output of the DAC is then amplified by the preamplifiers before the comparator makes a decision.

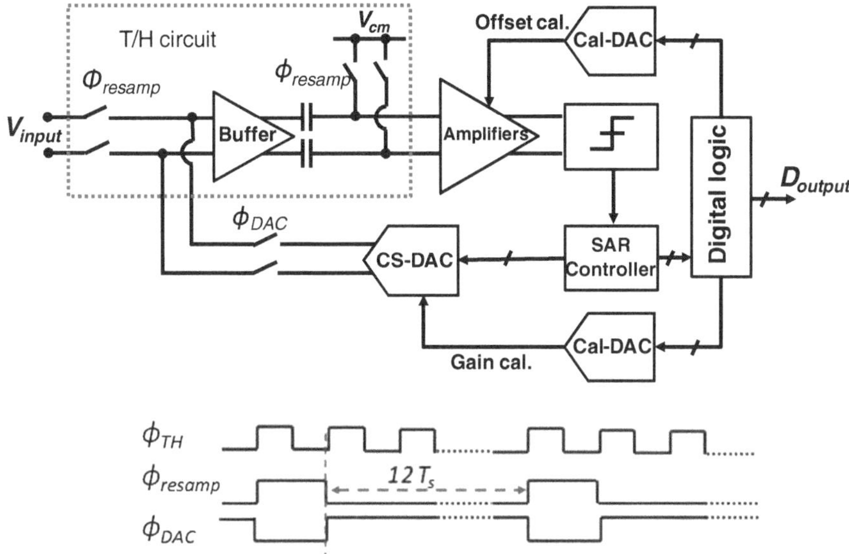

Fig. 4.26 Block diagram of the SAR ADC unit and its timing diagram [22]

The SAR-ADC units use a redundancy algorithm (sub-radix-2) to improve the conversion rate, resolving 11 b in 12 clock cycles excluding the sampling phase [21, 22].

Two calibration DACs are also included in the SAR ADC unit and they are controlled by an on-chip foreground calibration circuit to minimize the offset and gain mismatches among interleaved ADC units.

4.3.2.4 Range-Detection Circuits

Figure 4.27 shows the range detector which is a two-level Flash ADC. It consists of two dynamic comparators each preceded by three cascaded amplifiers as preamplifiers and a source follower buffer. Its input is connected directly to the front-end T/H. The input signal sampled by the front-end T/H is buffered by the source follower buffer and then compared with two on-chip generated reference voltages; the output of the detector is delayed to match the latency of the sub-ADC before it controls the output MUX. The range detector is designed to have 11 b resolution and operates at 1 GS/s.

There are a few design considerations for the range detection circuit. Firstly, the kickback noise to the sampling node, due to the fast regeneration and the reset of the comparator, needs to be minimized. As shown in Fig. 4.27, the input of the detection circuit is always connected to the output of the front-end T/H and operated at the same speed as the front-end T/H (1 GS/s); any disturbance to the sampling node results in signal distortion. Secondly, the input signal levels need to

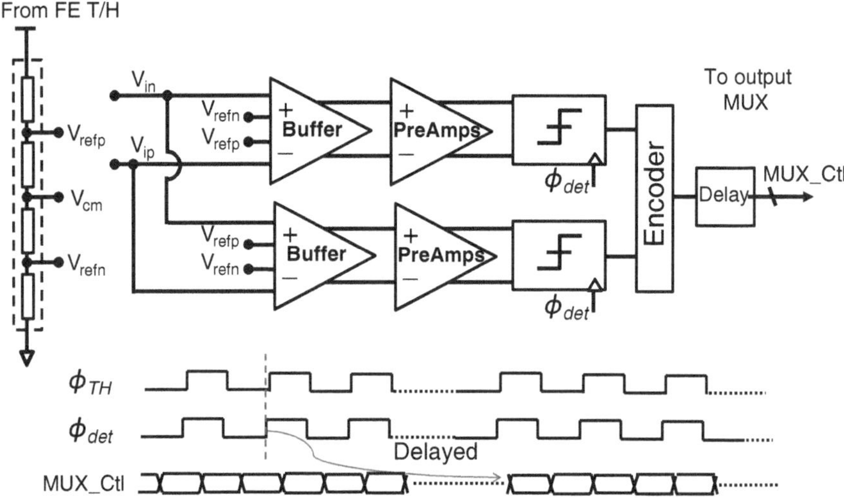

Fig. 4.27 Block diagram of the range detection circuit and its timing diagram

be detected with enough accuracy to avoid SNDR degradation of the reconstructed output signal. Thirdly, the output of the detection circuit needs to match the latency of the output of the sub-ADC, as it is used to control the output MUX that combines the outputs of two sub-ADCs.

Figure 4.28 shows the schematics of the range detection circuit block. The preamplifiers together with a dynamic latch are optimized to achieve fast-decision and low input referred noise as well as low kick-back noise. A source follower buffer is included for the purpose of buffering the output of the front-end T/H and reference voltages and to provide a low impedance node to minimize the kickback noise to the T/H and reference ladders.

4.3.2.5 Output MUX

The output MUX is designed to facilitate testing and reduce the number of pins needed to send out the data from the chip. It has three modes of operation: sending out the output data of either the main or auxiliary sub-ADC at 1 GS/s; multiplexing two sub-ADC outputs and sending out the combined output data at 2 GS/s, combining the outputs of two sub-ADCs based on the decision of the detection channel and sending them out at 1 GS/s. This allows having the main and auxiliary sub-ADC output and detection channel output fully available outside which is very important for evaluating each individual sub-ADC and detection channel.

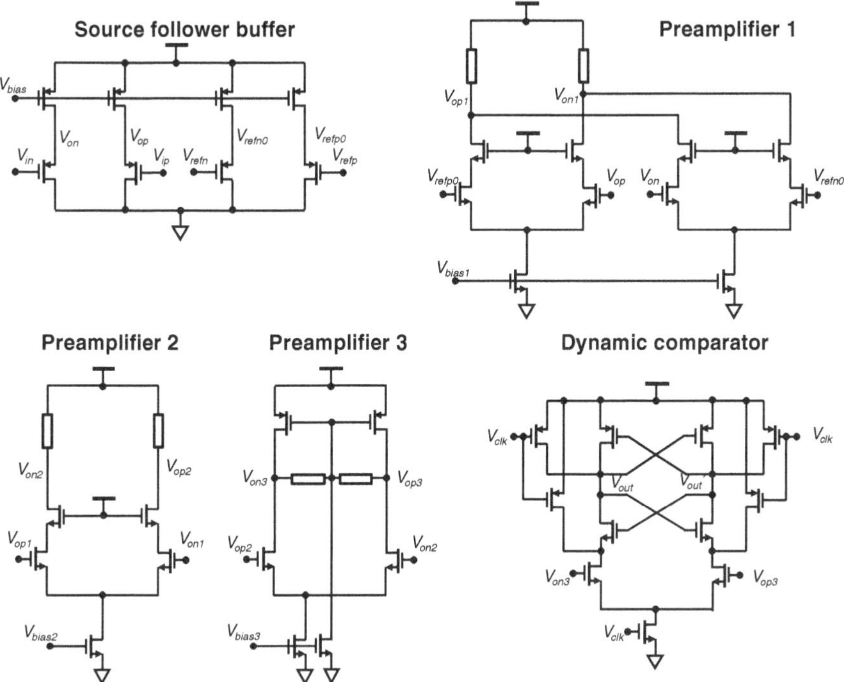

Fig. 4.28 Schematics of the range detector circuit block

4.3.3 Layout and Test-Chip Implementation

As this chip contains both sensitive analog circuitry and "noisy" digital circuitry on a single die, parallel sub-ADCs, and time-interleaved ADC units, a careful layout planning is necessary in order to reduce the performance degradation due to crosstalk and mismatch between channels. The layout floorplan of the ADC is shown in Fig. 4.29. In general, the sensitive analog part of the chip is kept away from the digital part; they are properly shielded and use separate voltage supplies to avoid potential crosstalk issues.

The placement of the major blocks and signal routing is as follows. The range scaling block is placed close to the bond pad on the bottom left. Two front-end T/Hs are located on the left. The clock generation and distribution network are located between the front-end T/Hs. Two parallel sub-ADCs, each consisting of 16 SAR-ADC units, are located in the center of the layout. The digital logic block and the output MUX are on the right. The global biasing and reference blocks are on the bottom. By designing the layout of the sub-ADC with a small width and long length, the distance between two the front-end T/H is minimized which allows lowering the power and skew of the full speed clock distribution. The analog input signal V_{in} and the analog power supplies enter the chip from the bottom, the external clock V_{clk} is

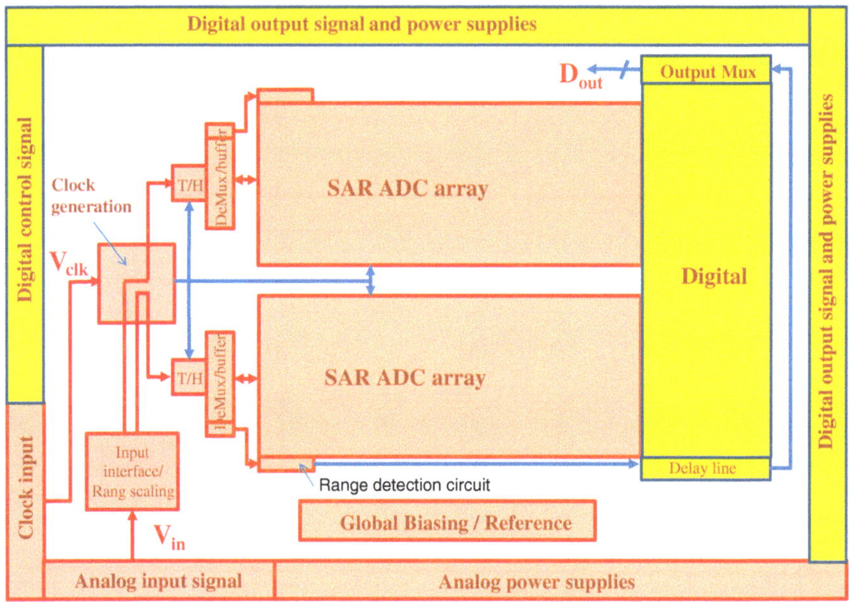

Fig. 4.29 The layout floorplan of the IC

routed from the left, while the digital outputs and power supplies are routed to/from the top and the right. The remaining area of the chip is filled with decoupling capacitance consisting of both CMOS capacitors and metal fringe capacitors.

The chip has 88 pads and includes a pad-ring with ESD protections. The pad ring is split into two separated parts, one for analog circuitry and one for digital circuitry. The purpose was to separate the power supply lines from digital and analog sections, in order to minimize crosstalk from the digital circuitry to the analog circuitry.

The prototype ADC has been fabricated in TSMC 65 nm LP CMOS. The chip photograph and its bonding diagram are shown in Fig. 4.30. The chip has a total area of 2.5 × 2.2 mm^2 including pads and the core area of the ADC is 2.3 × 1.8 mm^2. The drawn shapes highlight the location of the major blocks of this ADC in the chip. The chip is assembled in a HVQFN (Heatsink Very-thin Quad Flat-pack No-leads) package. In this package, all ground lines are down-bonded to the exposed die pad to reduce parasitic inductance.

4.3.4 Measurement Setup

The measurement setup used to gather measurement data for the prototype ADC is shown in Fig. 4.31, and the test equipment models are listed in Table 4.2. The test

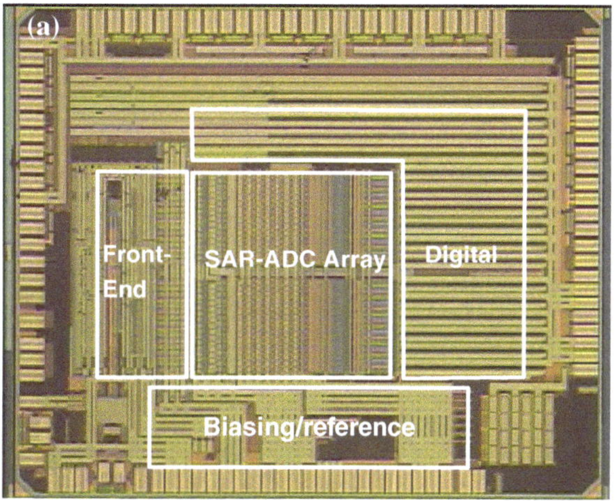

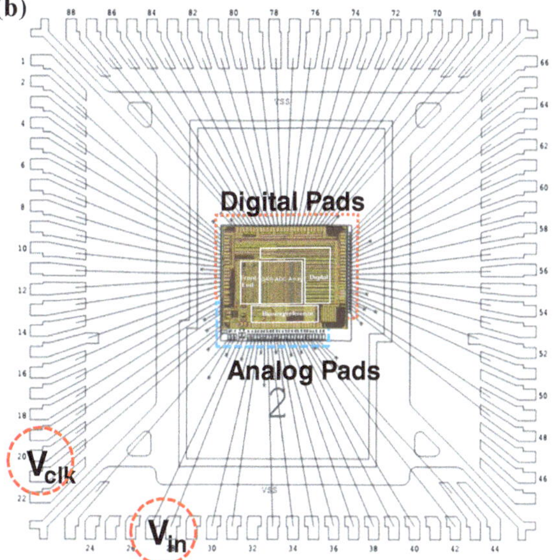

Fig. 4.30 **a** Die photograph of the IC and **b** package bonding diagram

setup consists of the device under test (DUT), the printed circuit board (PCB), several power supplies, signal generator for the input signal, clock generators for the ADC, USB to I^2C adapter for the control interface, data capture equipment, and a computer that processes the ADC output data.

In order to measure the performance of the ADC accurately, the measurement setup requires careful design to keep the nonidealities of the equipments and environment below the precision level of the chip. The chip is held with a high performance semi-customized test socket for HVQFN88 package to improve the connections between the chip and the PCB. The PCB provides various interfaces for the chip with the signal/clock generators, power voltage supplies, and the data

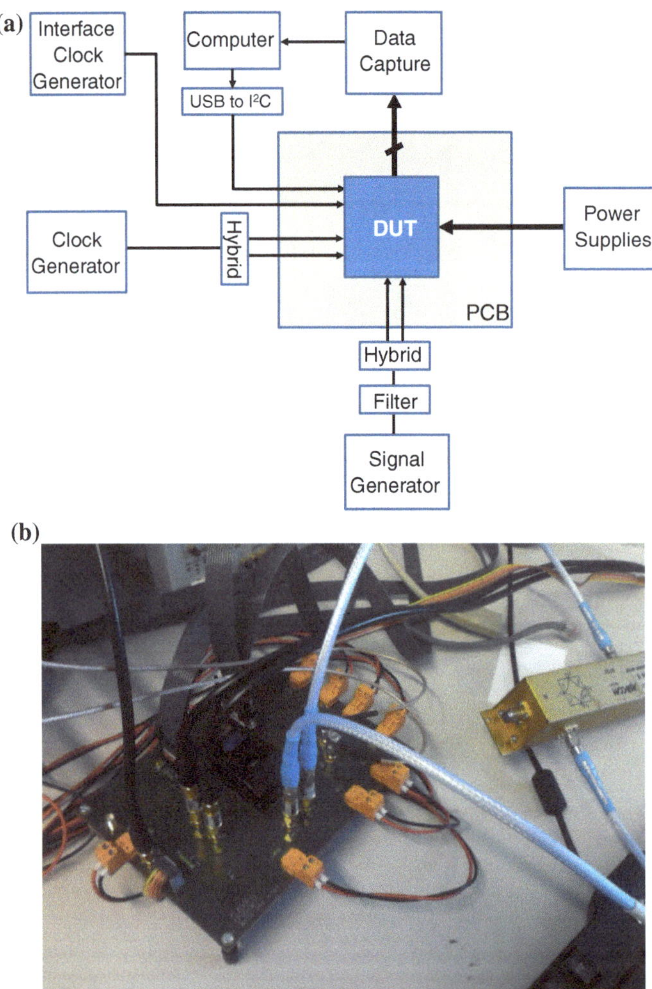

Fig. 4.31 **a** Measurement setup diagram, **b** PCB photograph

Table 4.2 Test equipments used in the measurement setup

Function	Name of test equipment
Signal generator (single sinusoid)	Marconi 2042 (10 kHz–5.4 GHz)
Signal generator (multi-carrier)	Agilent 81180A Arbitrary Waveform Generator
Clock generator	Anritsu 69177B (10 MHz–50 GHz)
Interface clock generator	Agilent 33220A 20 MHz Waveform Generator
Hybrid	M/A-COM 96341 (2-2000 MHz)
Data capture	Tektronix Logic Analyzers TLA7012
Power supplies	Agilent E3631A

capture equipment. The PCB is made with four layers including two layers for interconnections and two layers dedicated for ground and power distribution. Metal tracks on the PCB for input/output signals and clock are designed as symmetric transmission lines with 50 Ω impedance. Since the chip requires a differential clock, the single-ended output signal of the signal generator is converted to a differential signal by an off-board hybrid. An off-board tunable bandpass filter is used to filter out the undesired harmonics and tones from the testing signal before it goes into the chip. A similar option was also used for the input signal, as finding a single on-board transformer that meets the performance requirement for the large range of input frequencies is challenging. Customized software was used to program the chip control registers through the I^2C interface. The output data of the ADC is captured by the Tektronix TLA7012 logic analyzer which is able to capture data up to 2.8 GS/s and is sent to the computer for further analysis [43].

4.3.5 Experimental Results

This section presents the ADC experimental results. Various test signals are applied to the input of the ADC and its digital outputs are analyzed in the frequency domain using Discrete Fourier transform (DFT). Firstly, the dynamic performance of the sub-ADC is characterized with a single sinusoid input signal. Standard ADC performance metrics, such as SNDR, SFDR, SNR, THD, are shown with respect to input signal frequencies. Secondly, the parallel-sampling ADC is tested with a multi-carrier signal and its performance is compared with that of its sub-ADC which has state-of-the-art performance.

4.3.5.1 Performance of the ADC for a Single Sinusoid Signal

To verify the performance of the ADC across different input frequencies, the input signal frequency was swept from 19 MHz to 3 GHz with the ADC operating at 1 GS/s. The range of input signal frequencies in this measurement was limited by the bandwidth of the hybrid that converts the single-ended input signal from the signal generator to a differential signal. The measured SNR, SNDR, SFDR and harmonic distortions of the sub-ADC as a function of the input frequencies are shown in Figs. 4.32 and 4.33. The SNDR is 54.5 dB at low frequencies which is limited by thermal noise as it was designed, and stays above 50 dB up to 3 GHz thanks to low clock jitter and high frequency linearity. The SFDR is 79 dB at low frequencies, and it stays above 65 dB up to 1.5 GHz and maintains above 55 dB up to 3 GHz. The RMS jitter of the sampling clock is estimated to be around 100 fs which include the jitter from the external clock source. The measured ERBW is larger than 2.5 GHz. These measurements were done with both the sub-ADCs and detection path active, so non-idealities due the extra loading and disturbance to the ADC front-end are also included in the measurement results mentioned above.

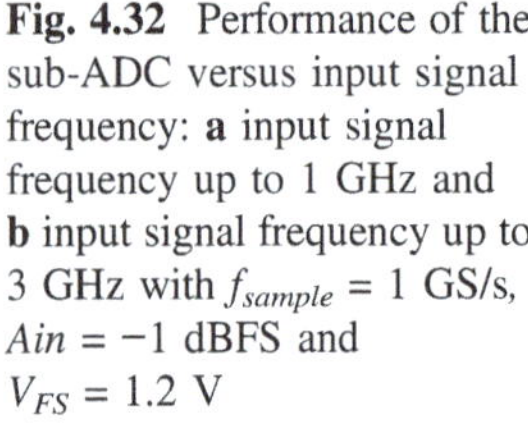

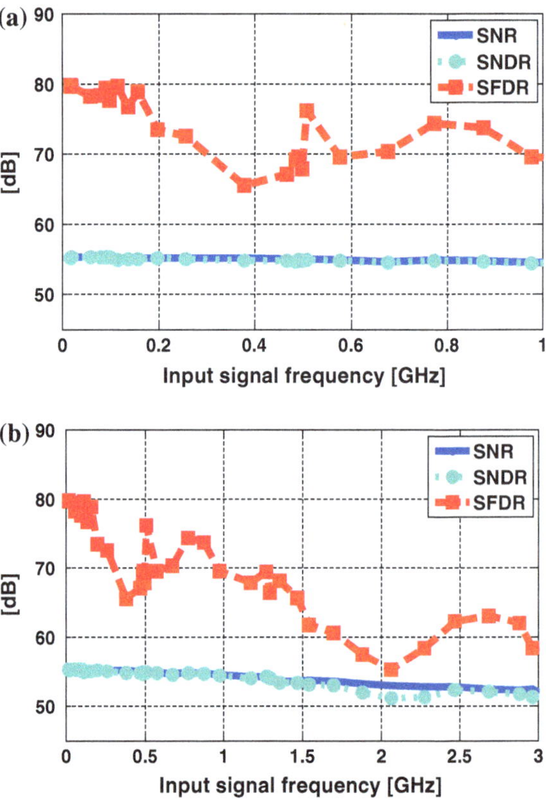

Fig. 4.32 Performance of the sub-ADC versus input signal frequency: **a** input signal frequency up to 1 GHz and **b** input signal frequency up to 3 GHz with f_{sample} = 1 GS/s, $Ain = -1$ dBFS and V_{FS} = 1.2 V

Figure 4.33 shows the different harmonics (HD2 to HD7) as a function of the input signal frequency. The SFDR is mostly dominated by the 3rd harmonic for input signal frequencies below 1.5 GHz. When the input signal frequencies are above 1.5 GHz (beyond the third Nyquist zone), the SFDR is mostly dominated by the second harmonic. Possible causes of the second harmonic higher than the third harmonic are: firstly, the phase imbalance of the off-chip hybrid that generates the differential input signals for testing the chip (the manufacturer specifies a maximum phase imbalance of 7° in the frequency range of 1–2 GHz [44]); secondly, the unequal length of the bondwires connecting the pins of package and the pads of the die for the differential input signal which is a design flaw (it can be observed in Fig. 4.30). These assumptions are verified by circuit simulations and they can be improved in the next design without much difficulty by using a wideband hybrid with smaller phase imbalance and optimizing the location of the corresponding pads. Nevertheless, the sub-ADC still shows better SNDR for sub-sampling input signals at around 3 GHz compared to GS/s ADCs published up to year 2014 in ISSCC, VLSI, ESSCIRC and CICC conferences [21–35].

The measured output spectra of the sub-ADC (main) for single sinusoid input signals with several frequencies are shown in Fig. 4.34. The ADC operates at 1 GS/

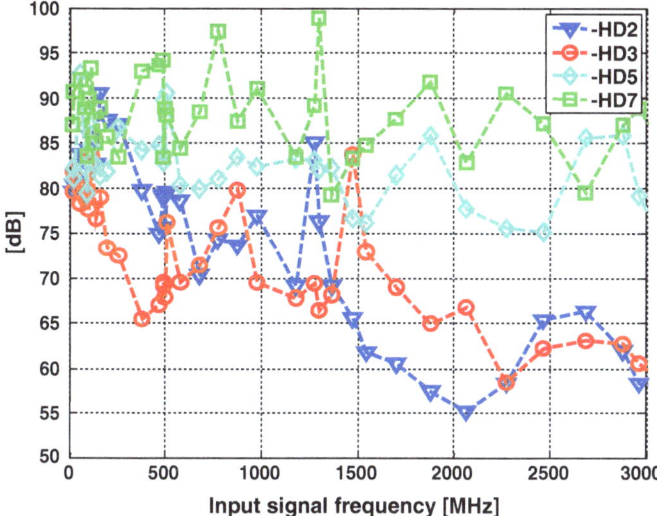

Fig. 4.33 Harmonics versus input signal frequency with f_{sample} = 1 GS/s, Ain = −1 dBFS and V_{FS} = 1.2 V

s and the input signal amplitude was kept at about −1 dBFS in the measurement. For input signals with frequencies in the 1st Nyquist bandwidth of the ADC, spurs due to the non-idealities of the time-interleaved ADC and disturbance of the other sub-ADC and range detector are well below −80dBFS. High spectrum purity is also achieved for sub-sampling a signal beyond the ADC's Nyquist bandwidth; as shown in Fig. 4.33, all the non-harmonic spurs are below −75 dBFS and harmonics are below −55 dB with input frequency of around 3 GHz which is at the 6th Nyquist bandwidth of this ADC.

Figure 4.35 shows the SNDR of the parallel-sampling ADC with respect to the input signal amplitude, the full scale input range of sub-ADC and the proposed ADC with parallel-sampling architecture are $1.2V_{ppd}$ and $2.4V_{ppd}$ respectively. The measured SNDR increases with the increase of input signal amplitude until the ADC begins to saturate; then the SNDR falls off sharply. The proposed ADC achieves a similar peak SNDR of about 55 dB for a single-sinusoid input signal compared to that of its sub-ADC and shows 6 dB improvement on the ADC's dynamic range as the linear input range of the proposed ADC is two times larger than its sub-ADC. Figure 4.36 shows the measured ADC output waveforms and histograms when processing a single sinusoid input signal with about $2.4V_{ppd}$ amplitude. As explained in the previous chapter, the main sub-ADC is allowed to clip while the auxiliary sub-ADC is not. The clipped samples from the main ADC are replaced by corresponding samples from the auxiliary ADC after they are amplified digitally; hence the reconstructed signal (output of the proposed ADC) doesn't suffer from excessive clipping noise.

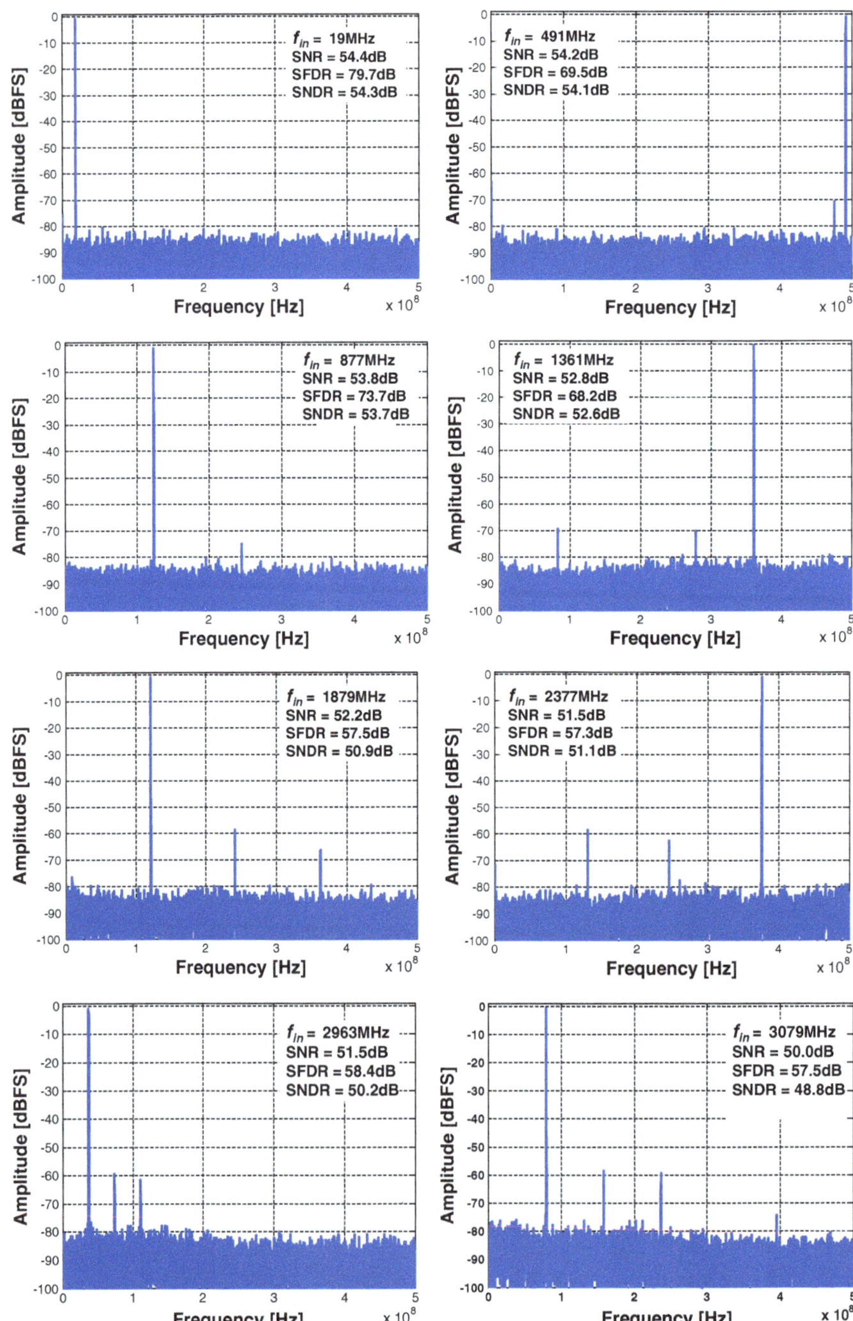

Fig. 4.34 Sub-ADC output spectrum for input signals with varies frequencies (f_{sample} = 1 GS/s)

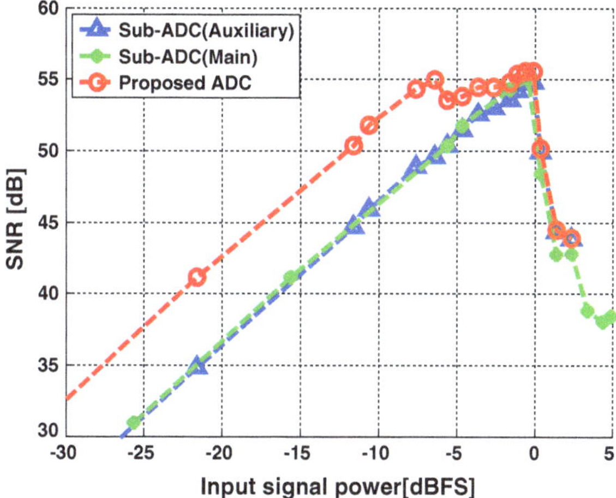

Fig. 4.35 SNDR versus input signal power for a single-sinusoid signal (*FS* refers to each ADC's full scale range)

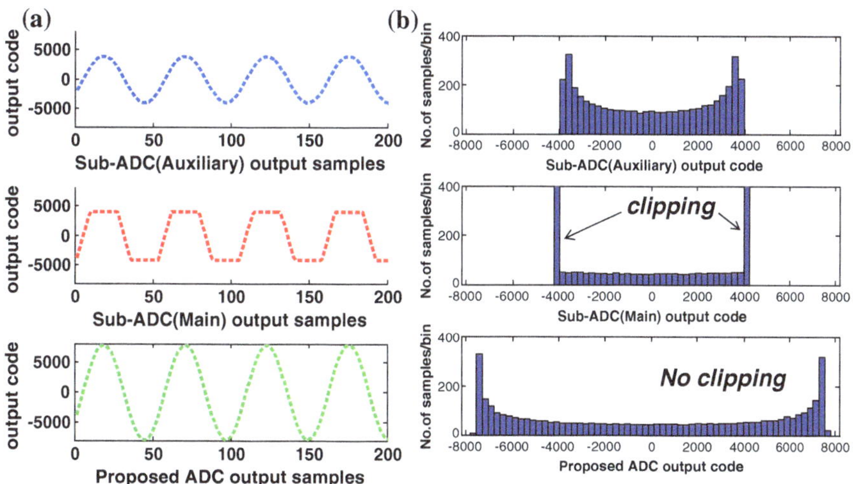

Fig. 4.36 Measured the ADC output **a** waveforms of a single sinusoid input signal and **b** their histograms

4.3.5.2 Performance of the ADC for Multi-Carrier Signals

In order to evaluate the performance of the proposed ADC with the parallel-sampling architecture for a broadband multi-carrier system, an NPR measurement is adopted which is more appropriate than the single or two-tone test as a measurement of the ADC performance for a broadband system as explained in Chap. 2.

A typical NPR measurement setup is shown in Fig. 4.37 [45]. The test signal should have a Gaussian-like amplitude distribution with a large PAPR, such as white noise or a multi-carrier signal. The testing signal used in this experiment is a multi-carrier signal. It is composed of 51 equal power channels having 6 MHz bandwidth and 8 MHz spacing; each channel uses 256QAM modulation and has random phase. All the channels are located within the first Nyquist band of the proposed ADC (up to 500 MHz), and a few empty channels are created on purpose for measuring the NPR. This test signal was generated by using MATLAB and it was downloaded to an arbitrary waveform generator (AWG). Due to the noise and linearity limitation of the AWG in the lab, the test signal generated by the AWG required extra filtering to get sufficient depth for the band-stop region for an accurate NPR measurement. A low-pass filter was used to filter out-of-band noise to prevent noise aliasing and band-stop filters were used as an option to improve the depth for the band-stop region. The NPR is calculated using Eq. (2.2) from the data captured, which was post-processed using MATLAB in a computer .

The measured NPR of the proposed ADC and the sub-ADC with respect to the input signal power is shown in Fig. 4.39. When the input signal amplitude is small, the NPR is dominated by the thermal noise floor of the ADC and it increases

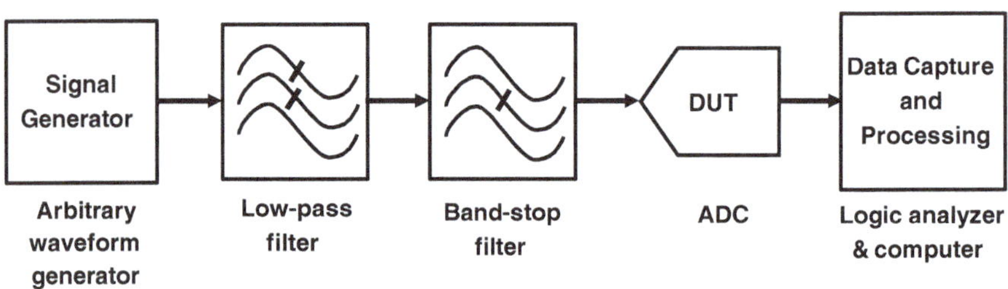

Fig. 4.37 A typical NPR measurement setup

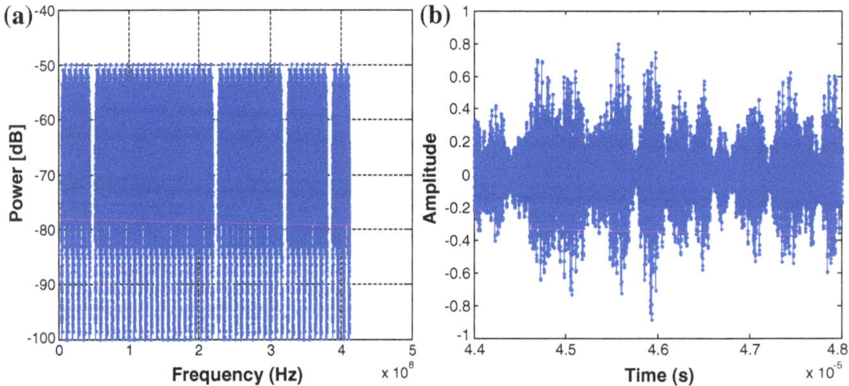

Fig. 4.38 **a** Spectrum of the test signal and **b** its time domain waveform

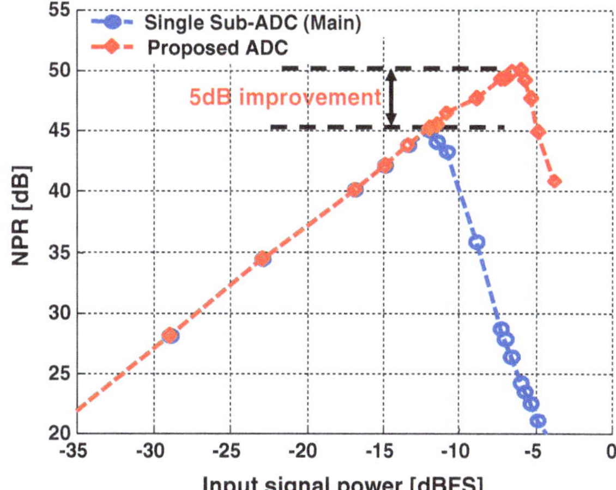

Fig. 4.39 NPR versus input signal power for a multi-carrier input signal (FS refers to the main sub-ADC's full scale range)

linearly with the input signal amplitude. When the instantaneous input signal amplitude exceeds the maximum input signal range of the ADC, clipping noise starts dominating the noise floor and results in a steep downward slope. As explained in the previous section, the input signal range of the proposed ADC is enlarged by 6 dB compared to that of each sub-ADC (from $1.2V_{ppd}$ to $2.4V_{ppd}$). The measured NPR of the proposed ADC is 5 dB more than what can be achieved by its sub-ADC. This demonstrates that the NPR of the sub-ADC which is a state-of-the-art design can be improved by 5 dB with the proposed architecture for converting multi-carrier signals.

Figure 4.40 shows the outputs of the proposed ADC and its sub-ADCs when the best NPR (50.6 dB) is measured. The main sub-ADC has much higher probability of being clipped compared to the auxiliary sub-ADC. By replacing the clipped samples from the main ADC by corresponding samples from the auxiliary ADC after they are amplified digitally, clipping noise of the reconstructed signal is minimized. In the reconstructed signals, majority of the samples (98.5 %) is taken from the main sub-ADC and only about 1.5 % of the samples are from the auxiliary sub-ADC, so the reconstructed signal shows higher signal power but without getting excessive clipping noise. Therefore, a 5 dB improvement in SNDR compared to its sub-ADC is shown in Fig. 4.39.

The spectra of the outputs of the proposed ADC and its sub-ADCs are shown in Fig. 4.41. The best NPR measured is about 50 dB, and it corresponds to 5 dB improvement compared to it is sub-ADC. As the bandwidth of the unused channels (bandstop regions) has the same bandwidth as the modulated channel, the measured NPR is also equal to the SNDR of a modulated channel in this test setup.

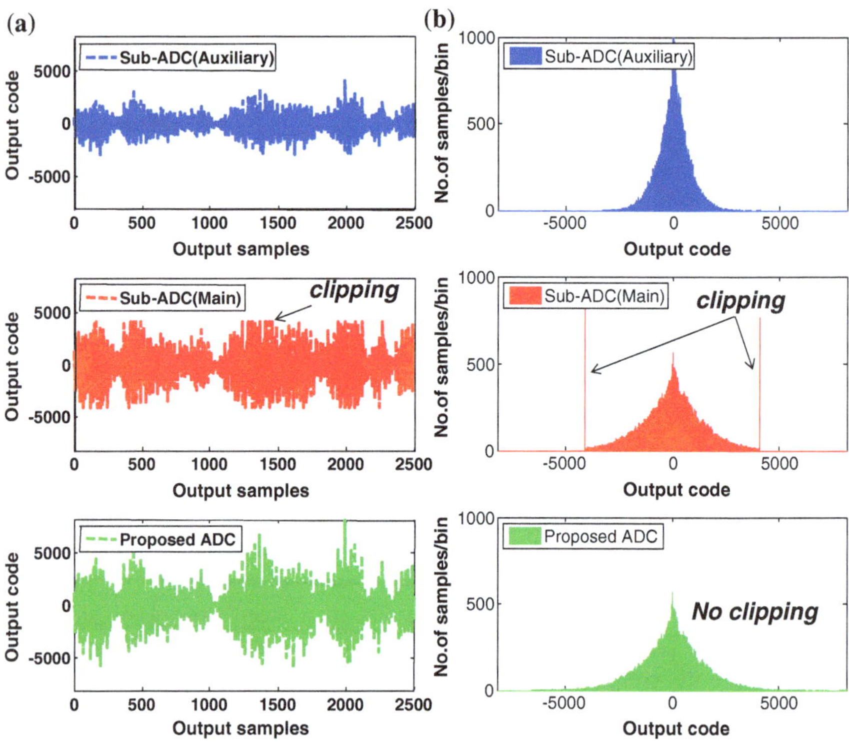

Fig. 4.40 Example of the measured ADC outputs: **a** waveforms of a multi-carrier signal and **b** their histograms

4.3.6 Performance Summary

This section summarizes the measured performance of the prototype IC as shown in Table 4.3. With a sampling rate of 1 GHz, the sub-ADC achieves an SNDR of better than 54 dB for input frequencies up to Nyquist frequency measured by a single sinusoid ($A_{in} = -1$ dBFS). The NPR of the prototype ADC with the parallel-sampling architecture is improved by 5 dB compared to its sub-ADCs when digitizing multi-carrier signals with a large PAPR and with a bell-shaped amplitude probability distribution. This improvement is achieved at less than half the cost in power and area compared to the conventional approach of using larger devices to lower the thermal noise power as explained in Chap. 3. The chip is implemented in 65 nm LP CMOS and consumes in total 350 mW at 1 GS/s including clock from 1. 2 to 2.5 V supplies.

The power consumption breakdown is shown in Fig. 4.42. Most of the power is consumed by the DACs in the SAR ADC units (34 %), and the source follower buffers in the frontend sampling and demultiplexing network (23 %). The rest of the

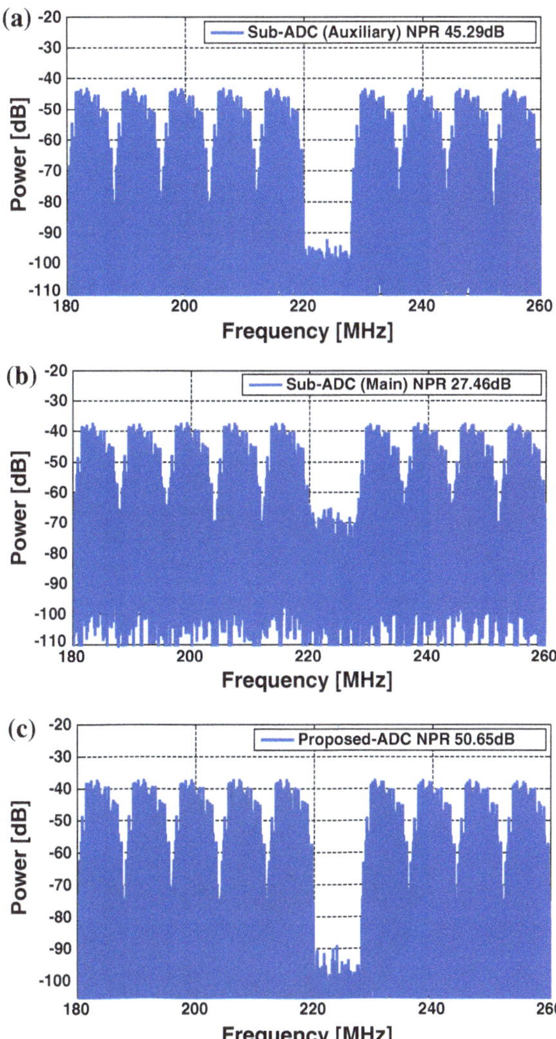

Fig. 4.41 The spectra of the outputs where the NPR is measured

power is divided among clock generation and distribution circuits (13 %), biasing and reference generation circuits (10 %), digital logics (10 %) and range detection circuit (2 %).

4.3.7 Comparison with State-of-the-Art

Table 4.4 lists ADCs with a sample rate of at least 1 GS/s and SNDR greater than 48 dB measured at the Nyquist frequency published at the International

Table 4.3 Performance summary

Process	65 nm LP CMOS
Sampling rate	1 GS/s
SFDR	79 dB @ 18 MHz;
	>65 dB up to 1 GHz;
	>55 dB up to 3 GHz
SNDR (single sinusoid signal)	54 dB @ Nyquist, single Sub-ADC
	>50 dB @ 3 GHz, single Sub-ADC
NPR/channel SNDR (multi-carrier signal)	45 dB, single Sub-ADC
	50 dB, proposed architecture
Jitter	<100f_s
Input sampling cap	200fF
Input signal range	2.4V_{ppd} (sub-ADC 1.2V_{ppd})
Supply voltages	1.2V and 2.5 V
Power	350 mW in total (including on-chip reference generation, clock buffers and biasing circuits, excluding output buffers);
Chip area	2.3 × 1.8 mm^2 (excluding pads)

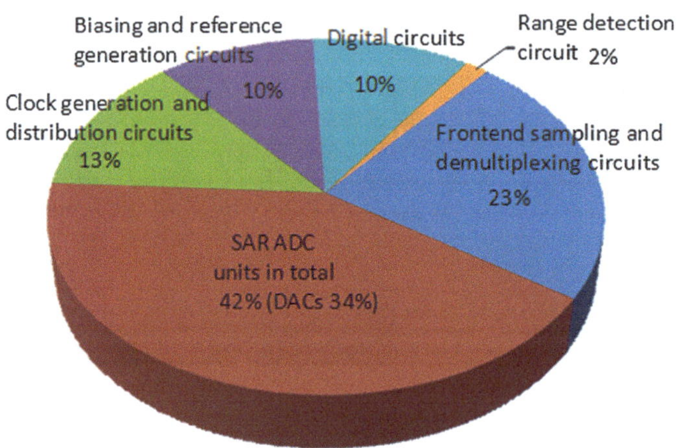

Fig. 4.42 Power consumption breakdown of the prototype ADC

Solid-State Circuits Conference (ISSCC), the VLSI Circuit Symposium (VLSI), the IEEE Custom Integrated Circuits Conference (CICC) and the European Solid-State Circuits Conference (ESSCIRC) up to year 2014. There are only 15 published ADCs in this performance range (including the work described in this book), these ADCs represent the current state-of-the-art in terms of speed and accuracy. The measured performance of the prototype ADC (both of the parallel-sampling ADC

Table 4.4 Selected data for performance comparison (All ADCs with $f_s \geq 1$ GHz and SNDR ≥ 48 dB published at ISSCC, VLSI, CICC, and ESSCIRC conferences up to year 2014)

Reference	Architecture	Technology	SNR (dB)	NPR (dB)	SNDR (dB)	P (mW)	f_s (GHz)	P/f_{snyq} (pj)	FOM_2 (dB)
[24] VLSI 2007	TI-SAR	130 nm	48.1		48.1	168	1.35	1.2E + 02	144.1
[25] VLSI 2011	TI-Pipeline	40 nm	60.0		51.0	500	3	1.7E + 02	145.8
[28] VLSI 2012	Pipeline	65 nm			52.4	33	1	3.3E + 01	154.2
[29] VLSI 2012	TI-SAR	65 nm	50.0		48.2	45	2.8	1.6E + 01	153.2
[32] VLSI 2013	TI-Pipeline	28 nm	61.0		57.0	500	5.4	9.3E + 01	154.3
[23] ISSCC 2006	TI-Pipeline	130 nm	58.0		52.0	250	1	2.5E + 02	145.0
[26] ISSCC 2011	TI-SAR	65 nm	52.0		48.5	480	2.6	1.8E + 02	142.8
[27] ISSCC 2011	TI-Pipeline	SiGe BiCMOS	59.0		59.0	575	1	5.8E + 02	148.4
[22] ISSCC 2013	TI-SAR	65 nm	52.0		50.0	795	3.6	2.2E + 02	143.5
[31] ISSCC 2013	TI-Pipeline	13 um BiCMOS	61.0		61.0	23900	2.5	9.6E + 03	138.2
[36] ISSCC 2014	TI-SAR	65 nm			51.4	19	1	1.9E + 01	155.6
[35] ISSCC 2014	TI-SAR	40 nm			48.0	93	1.62	5.7E + 01	147.4
[34] ISSCC 2014	Pipeline	65 nm	69.0		68.0	1200	1	1.2E + 03	154.2
[30] CICC2012	Pipeline	40 nm	58.0	46.0	52.0	280	2.1	1.3E + 02	147.7
[33] this work (Single sub-ADC)	TI-SAR	65 nm	54.2	45.3	**54.1**	211[b]	1	**2.1E + 02**	**147.8**
[33] this work (Parallel- sampling ADC)	TI-SAR	65 nm		50.6	**59**[a]	350[b]	1	**3.5E + 02**	**150.5**

[a]The proposed ADC shows 5.3 dB improvement in NPR in the measurement compared to that of its sub-ADC which corresponds to a similar improvement in SNDR

[b]Power consumption including on-chip reference generation and biasing circuits

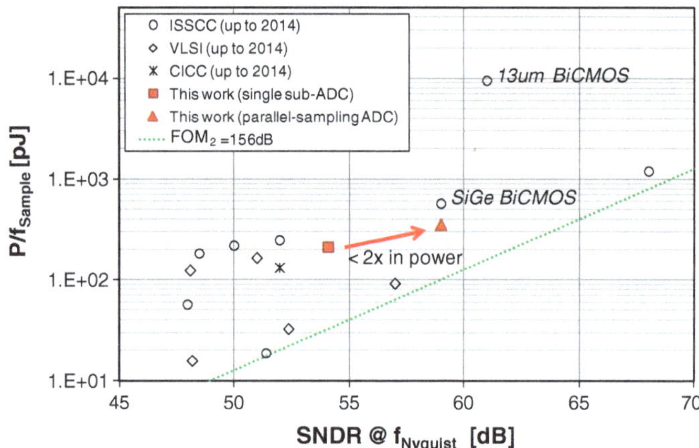

Fig. 4.43 Energy per conversion for ADCs with $f_s \geq 1$ GHz and SNDR ≥ 48 dB published in ISSCC, VLSI, CICC and ESSCIRC conferences up to 2014. (All these ADCs are in CMOS technology except two that are in BiCMOS technology as indicated.)

and its sub-ADC) is within this performance range. In Table 4.4, a comparison of these published ADCs with the prototype ADCs of this research work using the Schreier FOM (the FOM_2 mentioned in the Chap. 2 which is a suitable figure of merit for comparing noise-limited ADCs) and the conversion efficiency (power/conversion rate) is shown.

As validated by the measurements, the proposed ADC with parallel-sampling architecture is able to enhance the NPR and the channel SNDR of its sub-ADC by 5.3 dB for converting broadband multi-carrier signals. However, the performance of the proposed ADC for broadband multi-carrier signals is difficult to compare directly with other published ADCs as the reported performance of these publications is based on a single tone testing. In [18, 19], numerical simulations show that each additional bit of resolution corresponds to approximately 5.5 dB improvement in the NPR for an ideal ADC. Based on this finding, the proposed ADC has comparable performance with a conventional ADC having 59 dB SNDR for broadband multi-carrier signals. The FOM_2 and conversion efficiency of the prototype ADC with the parallel-sampling architecture are 150 dB and 350 pJ respectively. There are only 5 published ADCs in this performance range having a FOM_2 better than 150 dB, as shown in Table 4.4. Thanks to the parallel-sampling architecture, the FOM_2 of the parallel-sampling ADC is about 3 dB better than that of its sub-ADC but with less than 2 times in power consumption. In Fig. 4.43, the conversion efficiency of the ADCs listed in the Table 4.4 are plotted with respected to their SNDR values. Among these ADCs, there are only 3 published ADCs (two in BiCMOS and one in CMOS 65 nm) reporting an SNDR of more than 58 dB. The parallel-sampling ADC of this work shows a better conversion efficiency compared to that of these three ADCs, and only one of these ADCs (A.Ali, ISSCC2014, 1 GS/s and 68 dB SNDR) has a better FOM_2 but was published later than this work.

4.4 Conclusions

In this chapter, the parallel-sampling architecture presented in Chap. 3 was applied to two popular ADC architectures, the pipeline and time-interleaving SAR ADC architectures, to improve the ADC power efficiency for multi-carrier signals. The architecture study and circuit design of a parallel-sampling first stage for a 200 MS/s 12-b switched-capacitor pipeline ADC using TSMC 65 nm CMOS technology was first presented. A parallel-sampling frontend stage for a 4 GS/s 11 b time-interleaved ADC using GF 40 nm CMOS technology was presented next. It had been demonstrated by simulations that the parallel-sampling technique was able to enlarge the input signal range of the proposed ADCs, which is an effective way to improve the power efficiency of noise limited ADCs, without getting into excessive signal clipping. The simulated NPR of the proposed parallel sampling stages was improved by about 6 dB thanks to the additional parallel sampling auxiliary path and the statistical amplitude properties of the broadband multi-carrier signal. This improvement was achieved by less than half the power consumption and silicon area that would have been required when using the conventional approach with larger devices to lower the noise power.

This chapter also presented the design and experimental results of a prototype IC. This design has been implemented as a proof-of-concept of the parallel-sampling architecture for enhancing the SNDR of an ADC for broadband multi-carrier signals. The dynamic performance of the ADC was characterized with a single sinusoid input signal and a multi-carrier signal. When compared to ADCs with GHz sample rate of at least 1 GS/s and SNDR greater than 48 dB at Nyquist frequency published at ISSCC, VLSI, CICC, and ESSCIRC conferences up to year 2014, the Schreier FOM and the conversion efficiency (P/f_s) of the proposed ADC based on the parallel-sampling architecture and its sub-ADC are among the state-of-art. The parallel-sampling ADC of this work also shows a better conversion efficiency compared to ADCs with GHz sample rate and SNDR greater than 58 dB. This experiment demonstrates that the parallel-sampling architecture is able to enhance the performance of a state-of-art GS/s ADC (its sub-ADC, reused from [21, 22]) for broadband multi-carrier signals power efficiently.

References

1. Lin, Y., K. Doris, H. Hegt, and A.H.M. Van Roermund. 2012. An 11b pipeline ADC with parallel-sampling technique for converting multi-carrier signals. *IEEE Transactions on Circuits and Systems I: Regular Papers* 59(5): 906–914.
2. LTE-Advanced Physical Layer. http://www.3gpp.org/technologies/keywords-acronyms/97-lte-advanced. Accessed 15 June 2014.
3. IEEE Standard for Information Technology: Wireless LAN Medium Access Control (MAC) and Physical Layer (PHY) Specifications. *IEEE Std 802.11ac-2013*, 1–425, Dec. 2013.
4. Murmann, B. ADC Performance Survey 1997–2014. http://web.stanford.edu/~murmann/adcsurvey.html.

5. Gustavsson, M., J.J. Wikner, and N. Tan. 2000. *CMOS data converters for communications*, 2000th ed. Boston: Springer.
6. Chiu, Y. 2007. *Analysis and design of pipeline analog-to-digital converters*, 1st ed. Berlin: Springer.
7. Quinn, P.J., and A.H.M. Van Roermund. 2005. Design and optimization of multi-bit front-end stage and scaled back-end stages of pipelined ADCs. In *IEEE International Symposium on Circuits and Systems. ISCAS 2005*, 2005 (pp. 1964–1967, Vol. 3).
8. Abo, A.M., and P.R. Gray. 1999. A 1.5-V, 10-bit, 14.3-MS/s CMOS pipeline analog-to-digital converter. *IEEE Journal of Solid-State Circuits* 34(5): 599–606.
9. Ali, M.A., A. Morgan, C. Dillon, G. Patterson, S. Puckett, P. Bhoraskar, H. Dinc, M. Hensley, R. Stop, S. Bardsley, D. Lattimore, J. Bray, C. Speir, and R. Sneed. 2010. A 16-bit 250-MS/s IF sampling pipelined ADC with background calibration. *IEEE Journal of Solid-State Circuits* 45(12): 2602–2612.
10. Quinn, P.J., and A.H.M. van Roermund. 2007. Design criteria for cyclic and pipelined ADCs. In *Switched-capacitor techniques for high-accuracy filter and ADC design*. Springer, Netherlands, 2007, 165–192.
11. Murmann, B. 2012. Low-power pipelined A/D conversion. In *Analog circuit design*, ed. M. Steyaert, A. van Roermund, and A. Baschirotto, 19–38. Netherlands: Springer.
12. Kurose, D., T. Ito, T. Ueno, T. Yamaji, and T. Itakura. 2006. 55-mW 200-MSPS 10-bit pipeline ADCs for wireless receivers. *IEEE Journal of Solid-State Circuits* 41(7): 1589–1595.
13. Mehr, I., and L. Singer. 2000. A 55-mW, 10-bit, 40-Msample/s Nyquist-rate CMOS ADC. *IEEE Journal of Solid-State Circuits* 35(3): 318–325.
14. Chang, D.-Y. 2004. Design techniques for a pipelined ADC without using a front-end sample-and-hold amplifier. *IEEE Transactions on Circuits and Systems I: Regular Papers* 51(11): 2123–2132.
15. Sahoo, B.D., and B. Razavi. 2009. A 12-Bit 200-MHz CMOS ADC. *IEEE Journal of Solid-State Circuits* 44(9): 2366–2380.
16. Sumanen, L., M. Waltari, and K.A.I. Halonen. 2001. A 10-bit 200-MS/s CMOS parallel pipeline A/D converter. *IEEE Journal of Solid-State Circuits* 36(7): 1048–1055.
17. Choksi, O., and L.R. Carley. 2003. Analysis of switched-capacitor common-mode feedback circuit. *IEEE Transactions on Circuits and Systems II: Analog and Digital Signal Processing* 50(12): 906–917.
18. IEEE Standard for Terminology and Test Methods of Digital-to-Analog Converter Devices. *IEEE Std 1658-2011*, 1–126, Feb. 2012.
19. Kester, W.A. 2005. *Data Conversion Handbook*. Newnes, Burlington
20. DOCSIS 3.0 Physical Layer Interface Specification. http://www.cablelabs.com/specification/docsis-3-0-physical-layer-interface-specification/. Accessed 09 June 2014.
21. Doris, K., E. Janssen, C. Nani, A. Zanikopoulos, and G. Van Der Weide. 2011. A 480 mW 2.6 GS/s 10b time-interleaved ADC With 48.5 dB SNDR up to Nyquist in 65 nm CMOS. *IEEE Journal of Solid-State Circuits* 46(12): 2821–2833.
22. Janssen, E., K. Doris, A. Zanikopoulos, A. Murroni, G. van der Weide, Y. Lin, L. Alvado, F. Darthenay, and Y. Fregeais. 2013. An 11b 3.6GS/s time-interleaved SAR ADC in 65 nm CMOS. In *Solid-State Circuits Conference Digest of Technical Papers (ISSCC), 2013 IEEE International*, 464–465.
23. Gupta, S., M. Choi, M., Inerfield, and J. Wang. 2006. A 1GS/s 11b Time-Interleaved ADC in 0.13/spl mu/m CMOS. In *Solid-State Circuits Conference, 2006. ISSCC 2006. Digest of Technical Papers. IEEE International*, 2360–2369.
24. S.M. Louwsma, E.J.M. van Tuijl, M. Vertregt, and B. Nauta. 2007. "A 1.35 GS/s, 10b, 175 mW time-interleaved AD converter in 0.13 um CMOS." In *2007 IEEE Symposium on VLSI Circuits*, 62–63.
25. Chen, C.-Y., and J. Wu. 2011. A 12 b 3GS/s pipeline ADC with 500 mW and 0.4 mm^2 in 40 nm digital CMOS." In *2011 Symposium on VLSI Circuits (VLSIC)*, 120–121.
26. Doris, K., E. Janssen, C. Nani, A. Zanikopoulos, and G. Van Der Weide. 2011. "A 480 mW 2.6 GS/s 10 b 65 nm CMOS time-interleaved ADC with 48.5 dB SNDR up to Nyquist." In

Solid-State Circuits Conference Digest of Technical Papers (ISSCC), 2011 IEEE International, 180–182.

27. Payne, R., C. Sestok, W. Bright, M. El-Chammas, M. Corsi, D. Smith, and N. Tal. 2011. A 12 b 1 GS/s SiGe BiCMOS two-way time-interleaved pipeline ADC. In *Solid-State Circuits Conference Digest of Technical Papers (ISSCC), 2011 IEEE International*, 182–184
28. Sahoo, B.D., and B. Razavi. 2012. A 10-bit 1-GHz 33-mW CMOS ADC. In *2012 Symposium on VLSI Circuits (VLSIC)*, 30–31.
29. Stepanovic, D., and B. Nikolic. 2012. A 2.8GS/s 44.6 mW time-interleaved ADC achieving 50.9 dB SNDR and 3 dB effective resolution bandwidth of 1.5 GHz in 65 nm CMOS. In *2012 Symposium on VLSI Circuits (VLSIC)*, 84–85.
30. Wu, J., C.-Y. Chen, T. Li, W. Liu, L. He, S.S. Tsai, B. Chen, C.-S. Huang, J.-J. Hung, W.-T. Shih, H. Hung, S. Jaffe, L. Tan, and H. Vu. 2012. A 240mW 2.1GS/s 12b pipeline ADC using MDAC equalization. In *2012 IEEE Custom Integrated Circuits Conference (CICC)*, 1–4.
31. Setterberg, B., K. Poulton, S. Ray, D.J. Huber, V. Abramzon, G. Steinbach, J.P. Keane, B. Wuppermann, M. Clayson, M. Martin, R. Pasha, E. Peeters, A. Jacobs, F. Demarsin, A. Al-Adnani, and P. Brandt. 2013. A 14 b 2.5 GS/s 8-way-interleaved pipelined ADC with background calibration and digital dynamic linearity correction. In *Solid-State Circuits Conference Digest of Technical Papers (ISSCC), 2013 IEEE International*, 466–467.
32. Wu, J., A. Chou, C.-H. Yang, Y. Ding, Y.-J. Ko, S.-T. Lin, W. Liu, C.-M. Hsiao, M.-H. Hsieh, C.-C. Huang, J.-J. Hung, K.Y. Kim, M. Le, T. Li, W.-T. Shih, A. Shrivastava, Y.-C. Yang, C.-Y. Chen, and H.-S. Huang. 2013. A 5.4GS/s 12b 500 mW pipeline ADC in 28 nm CMOS. In *2013 Symposium on VLSI Circuits (VLSIC)*, C92–C93.
33. Lin, Y., K. Doris, E. Janssen, A. Zanikopoulos, A. Murroni, G. van der Weide, H. Hegt, and A. van Roermund. 2013. An 11b 1GS/s ADC with parallel sampling architecture to enhance SNDR for multi-carrier signals. In *2013 Proceedings of the ESSCIRC (ESSCIRC)*, 121–124.
34. Ali, A.M.A., H. Dinc, P. Bhoraskar, C. Dillon, S. Puckett, B. Gray, C. Speir, J. Lanford, D. Jarman, J. Brunsilius, P. Derounian, B. Jeffries, U. Mehta, M. McShea, and H.-Y. Lee. 2014. A 14b 1GS/s RF sampling pipelined ADC with background calibration. In *Solid-State Circuits Conference Digest of Technical Papers (ISSCC), 2014 IEEE International*, 482–483
35. Le Dortz, N., J.-P. Blanc, T. Simon, S. Verhaeren, E. Rouat, P. Urard, S. Le Tual, D. Goguet, C. Lelandais-Perrault, and P. Benabes. 2014. "A 1.62GS/s time-interleaved SAR ADC with digital background mismatch calibration achieving interleaving spurs below 70 dBFS. In *Solid-State Circuits Conference Digest of Technical Papers (ISSCC), 2014 IEEE International*, 386–388.
36. Lee, S., A.P. Chandrakasan, and H.-S. Lee. 2014. A 1GS/s 10 b 18.9 mW time-interleaved SAR ADC with background timing-skew calibration. In *Solid-State Circuits Conference Digest of Technical Papers (ISSCC), 2014 IEEE International*, 2014, 384–385.
37. Yang, J., T.L. Naing, and R.W. Brodersen. 2010. A 1 GS/s 6 Bit 6.7 mW successive approximation ADC using asynchronous processing. *IEEE Journal of Solid-State Circuits* 45 (8): 1469–1478.
38. Kurosawa, N., H. Kobayashi, K. Maruyama, H. Sugawara, and K. Kobayashi. 2001. Explicit analysis of channel mismatch effects in time-interleaved ADC systems. *IEEE Transactions on Circuits and Systems I: Fundamental Theory and Applications* 48(3): 261–271.
39. Tsai, T.-H., P.J. Hurst, and S.H. Lewis. 2006. Bandwidth mismatch and its correction in time-interleaved analog-to-digital converters. *IEEE Transactions on Circuits and Systems II: Express Briefs* 53(10): 1133–1137.
40. Louwsma, S., E. van Tuijl, and B. Nauta. 2011. Time-interleaved track and holds. In *Time-interleaved Analog-to-Digital Converters*, 5–38. Netherlands: Springer.
41. El-Chammas, M., and B. Murmann. 2012. *Background calibration of time-interleaved data converters*. New York: Springer

42. Razavi, B. 2013. Design considerations for interleaved ADCs. *IEEE Journal of solid-state circuits* 48(8): 1806–1817.
43. TLA7000 TLA7000 Logic Analyzers | Tektronix. http://www.tek.com/datasheet/tla7000/tla7000-series-data-sheet-0. Accessed 20 May 2014.
44. MACOM Hybrid Junction (2 MHz–2 GHz) Datasheet. http://cdn.macom.com/datasheets/H-9.pdf. Accessed 20 May 2014.
45. Marjorie Plisch, "Noise Power Ratio for the GSPS ADC. http://e2e.ti.com/cfs-file.ashx/__key/CommunityServer-Discussions-Components-Files/68/5775.ADC1xD1x00-NPR-for-E2E.pdf. Accessed 20 May 2014.

Chapter 5
Conclusions and Recommendations

Abstract This chapter presents the conclusions and gives recommendations for future research based on the insight gained during this study.

5.1 Conclusions

As most of the ADCs nowadays are designed for a specific application, exploiting signal and system properties which are a priori known offers opportunities for architecture innovation to further enhance the ADC performance in terms of better accuracy, higher speed, lower power consumption and smaller die size. Previous works that exploit specific properties of signals and systems for this purpose were studied and summarized in Chap. 2. We conclude that this so-called '*signal-aware*', '*system-aware*' or '*application-aware*' ADC design approach that exploits specific properties of signals and systems to enhance performance is promising.

Power reduction techniques at circuit and architecture level for thermal noise limited ADCs in advanced CMOS technologies were studied and summarized in Chap. 3. We conclude that the approach of improving the voltage efficiency of thermal noise limited ADCs is an effective way to enhance the ADC power efficiency for a desired SNR.

By exploiting the statistical amplitude properties of multi-carrier signal and combined with architecture innovation, a parallel-sampling architecture was introduced to enhance the voltage efficiency of the ADC for multi-carrier systems. The knowledge of the parallel-sampling architecture for multi-carrier signals is upgraded by analytical analysis and simulations. We conclude that the parallel-sampling architecture can improve significantly the linear input signal range of an ADC (even beyond the linear signal range of the ADC sampling stage and internal amplifier stages) and it can also be used to improve the SNCDR of an ADC for multi-carrier signals in a power and area efficient way. Four implementation options of the parallel-sampling ADC architecture were proposed, which include an implementation

Y. Lin et al., *Power-Efficient High-Speed Parallel-Sampling ADCs for Broadband Multi-carrier Systems*, Analog Circuits and Signal Processing,
DOI 10.1007/978-3-319-17680-2_5

option with run-time adaptation to allow sub-ADCs sharing between different signal paths to further improve ADC power efficiency compared to previous works [1, 2].

Architecture studies and circuit implementations of a parallel-sampling first stage for a 200MS/s 12 b switched-capacitor pipeline ADC using 65 nm CMOS technology and a parallel-sampling frontend stage for a 4GS/s 11 b time-interleaved ADC using 40 nm CMOS technology were presented in Chap. 4. From circuit simulations and analysis, we prove that the linear input signal range of the proposed ADCs with a parallel-sampling stage can be enlarged by a factor of two or even more compared to their sub-ADCs without getting excessive clipping distortion. Simulation results showed at least 6 dB improvement of the ADC's dynamic range and about 6 dB improvement of NPR for a multi-carrier signal. This improvement is achieved by less than half the power consumption and silicon area that would have been required when using the conventional approach with larger devices to lower the noise power, which proves the parallel-sampling ADC architecture can effectively improve the power efficiency of noise limited ADCs for multi-carrier signals.

A prototype IC was implemented to demonstrate the feasibility of designing a GHz sample rate and noise-limited ADC with the parallel-sampling architecture and the advantage of this architecture in enhancing the power efficiency of noise-limited ADCs for multi-carrier systems. The experimental IC contains two 1 GS/s 11 b sub-ADC with a parallel-sampling frontend and a run-time signal-range detection path. Experimental results show that each of its sub-ADCs achieves thermal-noise limited SNR of above 54 dB for input frequencies up to 1.5 GHz (third Nyquist zone) and the frontend stage of the ADC demonstrates state-of-the-art linearity performance (SFDR > 65 dB up to 1.5 GHz and >55 dB up to 3 GHz). Experimental results show that the dynamic range of the ADC is improved by 6 dB compared to its sub-ADC for both a single sinusoid signal and a broadband multi-carrier signal, and the NPR is improved by 5 dB compared when digitizing a multi-carrier signal with large crest factors at a cost of only about two times increase in power and area (compared to at least four times using the conventional approach with larger devices to lower the noise power). The experimental results are compared with prior art with similar performance (with sample rate at least 1GS/s and SNDR more than 48 dB), using the Schreier FOM and the conversion efficiency (P/f_s). This experiment proofs the feasibility of designing a parallel-sampling ADC in this performance range and the advantage of this architecture in enhancing the power efficiently of a state-of-art GS/s ADC (refers to its sub-ADC, reused from [5]) for a better performance for multi-carrier systems.

5.2 Recommendations for Future Research

Due to the demand and challenges observed, designing more advanced ADCs will still be an active research topic and the concept of exploiting specific signal and system properties to enhance ADC performance for a specific application will be useful to bridge the gap. We suggest several directions for further work based on the insight gained during this research:

- The parallel-sampling architecture can be extended to include signal conditioning blocks in front of the ADC. Signal conditioning blocks, such as analog filters, LNA and PGA, require a high dynamic range and high sensitivity simultaneously. They also consume significant power and occupy large chip area compared to the ADC. Previous works have demonstrated enhancement techniques for analog filters based on similar ideas [3, 4]. It would be an interesting research topic to combine these techniques to yield a better baseband solution.
- Other ADC performance enhancement techniques for specific signals and systems, as summarized in Chap. 2 can also be combined with the parallel-sampling ADC architecture to further enhance the performance of ADCs in terms of better accuracy, higher flexibility or lower power consumption for a specific application.
- The signal-to-thermal-noise-power ratio of the ADC for converting multi-carrier signals is improved with the parallel sampling technique, but this does not include the noise due to jitter of the clocking circuits. Reducing jitter noise for high sample rate ADC requires significant power. Future work could also investigate the feasibility of applying the signal/system aware design approach to develop enhancement techniques for clocking circuits for a specific application.
- Furthermore, the concept of developing enhancement techniques for ADCs based on a priori information of the signal and system can also be applied to DACs which can be an interesting research topic as well.

References

1. Gregers-Hansen, V., Brockett, S.M., and P.E. Cahill 2001. A stacked A-to-D converter for increased radar signal processor dynamic range. In *Proceedings of the 2001 IEEE Radar Conference, 2001*, pp. 169–174.
2. Doris, K. 2010. Data processing device comprising adc unit, U.S. Patent US20100085231 A1, 08 April 2010.
3. Palaskas, Y., Tsividis, Y., Prodanov, V., and V. Boccuzzi. 2004. A 'divide and conquer' technique for implementing wide dynamic range continuous-time filters. *IEEE Journal of Solid-State Circuits* 39(2): 297–307.
4. Maheshwari, V., Serdijn, W.A., Long, J.R., and J.J. Pekarik. 2010. A 34 dB SNDR instantaneously-companding baseband SC filter for 802.11a/g WLAN receivers. In *Solid-State Circuits Conference Digest of Technical Papers (ISSCC), 2010 IEEE International*, 2010, pp. 92–93.
5 Janssen, E., Doris, K., Zanikopoulos, A., Murroni, A., van der Weide, G., Lin, Y., Alvado, L., Darthenay, F., and Y. Fregeais. 2013. An 11b 3.6GS/s time interleaved SAR ADC in 65 nm CMOS. In *Solid State Circuits Conference Digest of Technical Papers (ISSCC), 2013 IEEE International*, pp. 464–465.

Appendix
A Dynamic Latched Comparator for Low Supply Voltage Applications

The comparator is a key building block for applications where digital information needs to be recovered from analog signals, such as analog-to-digital (A/D) converters, I/O data receivers, memory bit line detectors, etc. The trend of achieving both higher speed and lower power consumption in these applications makes dynamic latch comparators very attractive, as they achieve fast decisions by strong positive feedback [1] and have no static power consumption. This appendix presents a dynamic latched comparator suitable for applications with very low supply voltage [2].

Introduction

Two commonly used dynamic latched comparators, the 'StrongARM' and the 'Double-tail' latched comparators [3–4], are shown in Fig. A.1. The 'StrongARM' comparator consists of an input differential pair and two cross-coupled nMOS and pMOS pairs which are stacked on top of each other. It achieves a fast decision due to strong positive feedback enabled by two cross-coupled pairs, and a low input referred offset enabled by the input differential pair stage. As shown in Fig. A.1a, this comparator requires stacking multiple of transistors on top of each other ($S0$, $M1$–3) which demands quite a large voltage headroom and becomes problematic with the supply voltage scaling in advanced CMOS technologies. Furthermore, since the input pair is stacked with the pMOS and nMOS cross-coupled pairs, the current flowing through these cross-couple pairs is limited by the input common-mode voltage of the differential pair, hence the speed and offset of such a circuit are greatly dependent on its input common-mode voltage, which can be a problem for applications requiring a wide common-mode range, for example A/D converters.

In [4], a 'Double-tail' latched comparator with a separated input and cross-coupled stages was introduced to mitigate the drawback of the 'StrongARM'

Y. Lin et al., *Power-Efficient High-Speed Parallel-Sampling ADCs for Broadband Multi-carrier Systems*, Analog Circuits and Signal Processing,
DOI 10.1007/978-3-319-17680-2

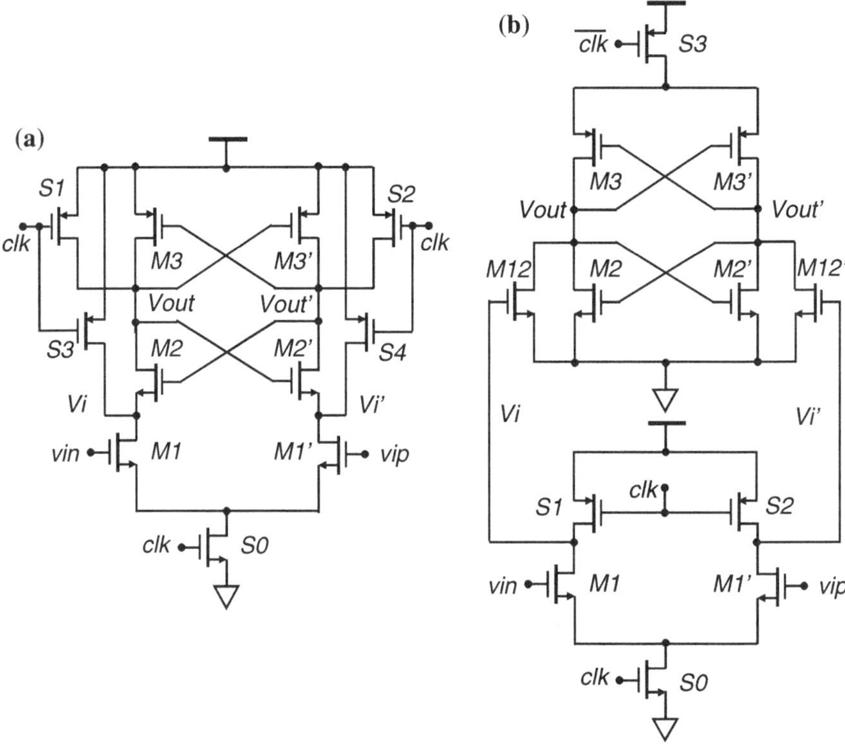

Fig. A.1 **a** the 'StrongARM' latched comparators; **b** the 'Double-tail' dual-tail dynamic comparator

latched comparators mentioned aboved, as shown in Fig. A.1b. This separation allows the input common-mode voltage and speed of the comparator to be optimatized independently, and allows it to operate at a lower supply voltage as well [4]. However, the stacking of two cross-coupled nMOS (*M*2 and *M*2′) and pMOS pairs (*M*3 and *M*3′) in this comparator still requires quite a large voltage headroom to accommodate two threshold voltages (V_T) plus two overdrive voltage ($V_{ds.sat}$) for the transistors to work in saturation. This limits the achievable speed of the comparator at low supply voltage or it may simply fail to work at very low supply voltage.

A novel dynamic latched comparator with a input stage and two separated cross-coupled pair stages was proposed in [2] as shown in Fig. A.2 and its linear time variant model in Fig. A.3. Compared to the previous works in [3–4], the major difference is that two cross-coupled pairs are placed in parallel in this comparator instead of stacking them on top of each other which is similar to the clock divider circuit shown in [5]. This circuit topology makes it suitable to work at very low supply voltage without compromising speed compared to [3–4]. It is a fully dynamic circuit without any static power consumption.

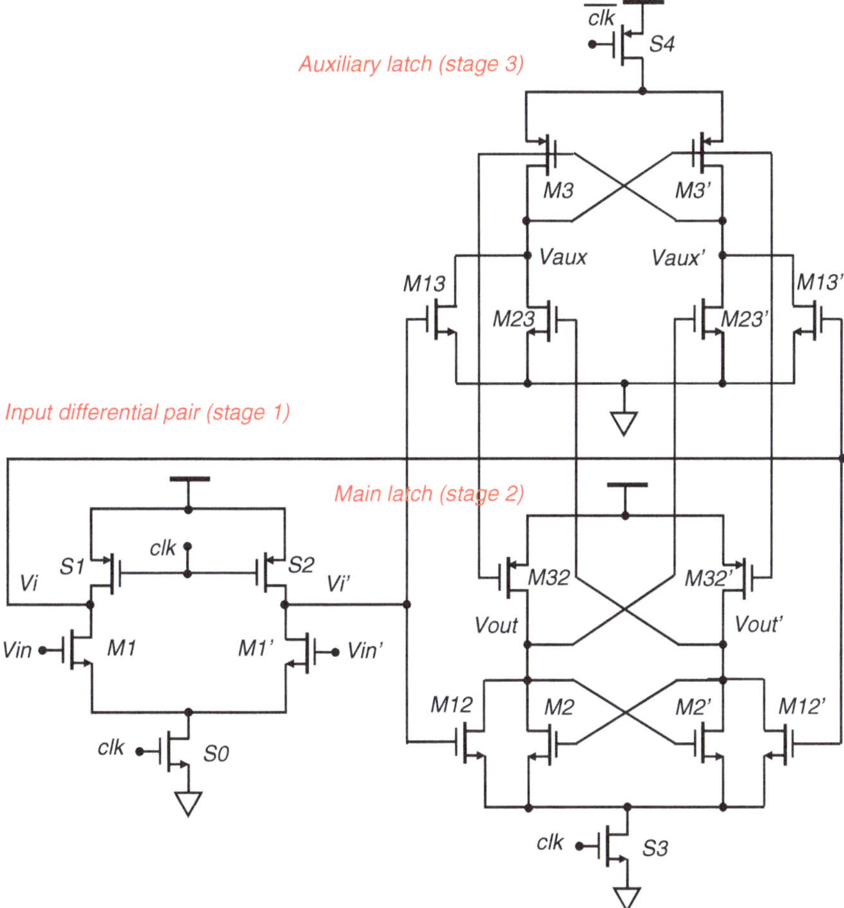

Fig. A.2 Schematic of the proposed dynamic latched comparator with parallel cross coupled pair stages

Circuit Description and Operation

The proposed comparator shown in Fig. A.2 consists of a differential pair input stage, two latch stages in parallel namely main stage and auxiliary stage, and some reset switches. Each of the latch stages consists of a cross-coupled pair and two input differential pairs. One of the input differential pairs ($M12/M12'$ or $M13/M13'$) of the latch stage passes the input signal for comparison, while the other one ($M32/M32'$ or $M23'/M23$) is controlled by the output of the other latch stage which forms another positive feedback loop for the signal besides the cross-coupled pairs during regeneration.

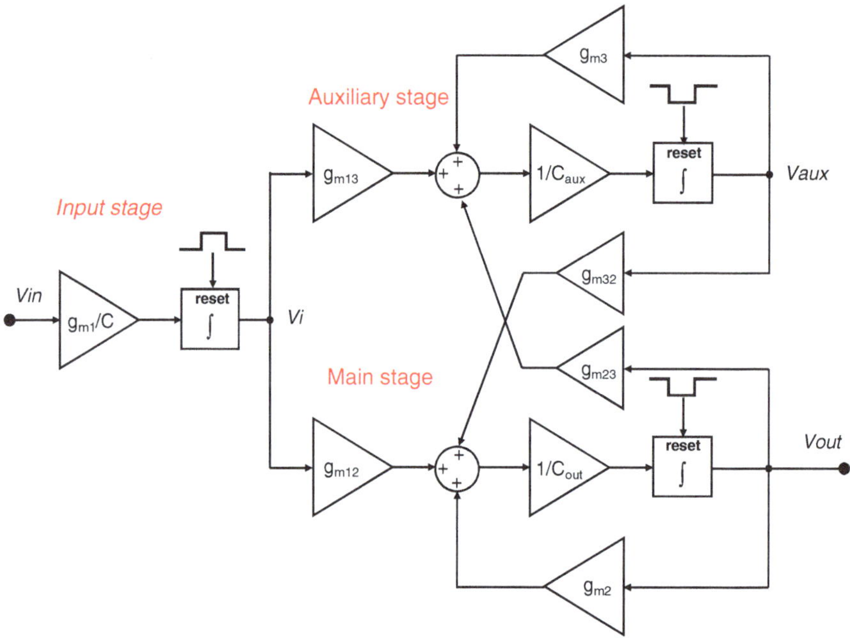

Fig. A.3 Linear time-variant model of the dynamic latched comparator with parallel cross coupled pair stages

Operation of this comparator is in principle similar to that of the 'StrongARM' and the 'Double-tail' latched comparators shown in Fig. A.1. As shown by the signal waveforms in Fig. A.4, it goes through a set of operating phases each cycle, namely: resetting, sampling, regeneration, and decision phase [6].

The comparator is in the resetting phase when the clock signal *clk* is low. The output nodes (*Vout*/*Vout′*) and internal nodes (*Vi*/*Vi′*) are precharged to the supply voltage by switches *S*1–*S*2, while nodes *Vaux*/*Vaux′* are discharged to ground, as it is shown in Fig. A.4. Switches *S*0 and *S*3–*S*4 are off to prevent static current consumption in this phase.

The sampling phase starts when *clk* is switched to high. The switches *S*1–*S*2 are turned off, the input differential pair (*M*1/*M*1′) discharges the nodes *Vi* and *Vi′* at a rate depending on the input signal difference (*Vin* − *Vin′*), hence an input dependent differential voltage Δvi is built up at the output of the input stage. The intermediate stages formed by *M*12/*M*12′ and *M*13/*M*13′ then pass this voltage difference to the nodes *Vaux*/*Vaux′* and *Vout*/*Vout′* in both the auxiliary and main latch stages to be used as an initial voltage for regeneration.

The regeneration phase begins when the cross-coupled transistors in the main stage and the auxiliary stage are turned on. The initial voltage difference built up at the nodes *Vaux*/*Vaux′* and *Vout*/*Vout′* during the sampling phase causes both the cross-coupled pairs to leave the unstable equilibrium state. Both the cross-coupled

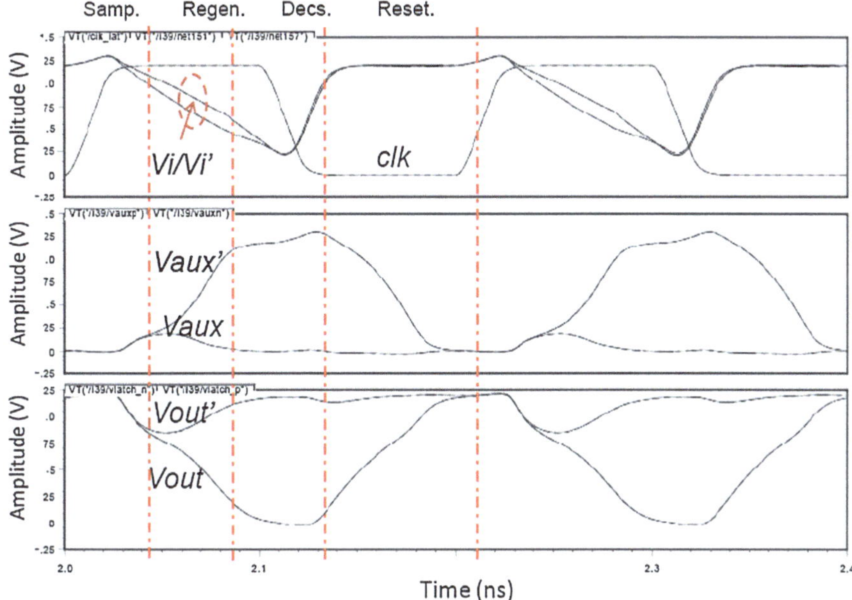

Fig. A.4 Simulated signal waveforms of the proposed comparator

pair stages ($M2/M2'$ and $M3/M3'$) regenerate the voltage difference via positive feedback in a parallel fashion. The output nodes of these parallel stages are also coupled by two g_m stages ($M23/M23'$ or $M32/M32'$) which forms another positive feedback loop to further enhance the regeneration speed, as will be detailed in the following analysis.

Small Signal Analysis and Design Considerations

The small signal model of the comparator in the sampling phase is shown in Fig. A.5a. Assuming the sampling phase lasts until t_0, the transfer function can be derived as follows:

$$\Delta vi(t_0) = \frac{1}{C_i}\int_0^{t_0} g_{m1} \cdot \Delta vin \cdot dt = \frac{g_{m1}}{C_i} \cdot \Delta vin \cdot t_0 \tag{A.1}$$

$$\Delta vo(t_0) = \frac{1}{C_{out}}\int_0^{t_0} g_{m12} \cdot \Delta vi \cdot dt = \frac{g_{m1} \cdot g_{m12}}{C_i \cdot C_{out}} \cdot \Delta vin \cdot t_0 \tag{A.2}$$

where Δvin is equal to the input signal difference $Vin - Vin'$, and $\Delta vi(t)$ and $\Delta vo(t)$ are equal to the voltage difference of the internal nodes $Vi - Vi'$ and output nodes $Vout - Vout'$ respectively, g_{m1} and g_{m12} are the transconductances of transistors $M1/M1'$ and $M12/M12'$, and C_i and C_{out} are capacitance at nodes Vi and $Vout$. Note that the resulting transfer function from Δvin to $\Delta vo(t)$ corresponds to two cascaded integrations.

The sampling time t_0 in (A.2) is proportional to the capacitor sizes (C_i and C_{out}) over their discharging current (I_d), therefore it is important to maximize the transconductance over discharging current (g_{m1}/I_{d1} and g_{m12}/I_{d2}) in order to have a high gain in this stage to reduce the input referred offset and noise voltage from the latch stages.

The regeneration phase can be approximately analyzed using a small signal model as shown in Fig. A.5b. In this equivalent circuit, g_{m2}, g_{m3}, g_{m23} and g_{m32} are the transconductances of $M2/M2'$, $M3/M3'$, $M23/M23'$ and $M32/M32'$ respectively. From this small signal model, using the *KCL* for the output nodes we can derive the following set of coupled first order differential equations as follows:

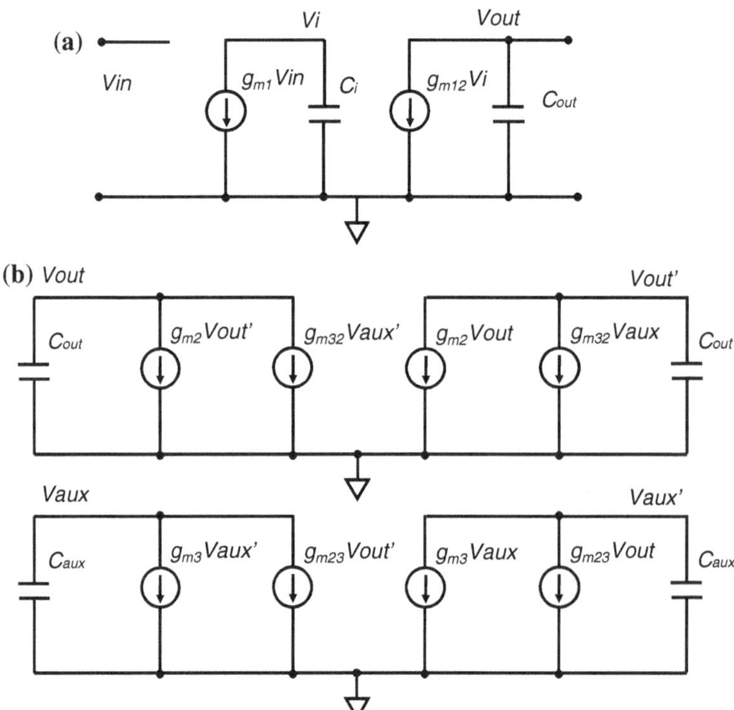

Fig. A.5 Small signal models of the proposed comparator in **a** sampling phase and **b** regeneration phase

$$\begin{cases} C_{out} \cdot \dfrac{dVout(t)}{dt} = -g_{m2} \cdot Vout'(t) - g_{m32} \cdot Vaux'(t) \\ C_{aux} \cdot \dfrac{dVaux(t)}{dt} = -g_{m3} \cdot Vaux'(t) - g_{m23} \cdot Vout'(t) \end{cases} \tag{A.3}$$

$$\begin{cases} Vaux(t) = -Vaux'(t) \\ Vout(t) = -Vout'(t) \end{cases} \tag{A.4}$$

Assuming $g_{m2} = g_{m3}$, $g_{m23} = g_{m32}$ and $C_{out} = C_{aux}$ for simplicity, the output differential voltage of the comparator $\Delta Vout(t)$ can be obtained by solving (A.3) and (A.4) with the initial condition of $Vaux(0) = Vout(0) = \Delta vo(t_0)/2$. $\Delta Vout(t)$ can be expressed by:

$$\begin{aligned} \Delta Vout(t) &= \Delta vo(t_0) \cdot \exp(\frac{g_{m2} + g_{m32}}{C_{out}} \cdot t) \\ &= \frac{g_{m1} \cdot g_{m12}}{C_i \cdot C_{out}} \cdot \Delta vin \cdot t_0 \cdot \exp(\frac{g_{m2} + g_{m32}}{C_{out}} \cdot t) \end{aligned} \tag{A.5}$$

From (A.5), we can observe that the regeneration speed depends on g_{m2}, g_{m32} and C_{out}. C_{out} is the output loading capacitance of the comparator which equals the sum of the external load C_{ext} and a parasitic component of transistors C_{lat}. As g_m is proportional to $W/L \cdot V_{gt}$, where W and L denote the width and length of a transistor and V_{gt} the effective gate-source overdrive voltage. C_{lat} is proportional to $W \cdot L$ which is mainly contributed by *M*2 and *M*23 in this comparator. By maximizing V_{gt} at the initial step of the regeneration phase, using minimum transistor length and taking into account C_{ext} to optimize the width of the transistors, an optimal regeneration speed can be achieved [7].

Comparison with Previous Works

The separation of the input stage and the cross-coupled pair stages keeps the same advantage as the 'Double-tail' latched comparator [4]. The additional intermediate stage (*M*12/*M*12′ *or M*13/*M*13′) provides additional shielding between the input and output of the comparator which results in less kickback noise. Besides that, the separation of the cross-coupled pair stages allows the latch stages to work at a lower supply voltage due to the reduction of the number of stacking transistors. The minimum supply voltage now required to assure the transistors in the latch stages to work in saturation is only one V_T plus two $V_{ds.sat}$. Therefore, the supply voltage required for the proposed comparator is at least one V_T lower than the previous works shown in Fig. A.1. The separation of the input stage and latch stages could result in extra power consumption as more nodes need to be charged and discharged, but it enables the comparator to operate at a much lower supply voltage without compromising speed.

Furthermore, when *clk* goes high, both the cross-coupled pairs of the proposed comparator become active at the same time with a large V_{gt} equals $V_{dd} - V_T$. While the cross-coupled pairs in both the comparators shown in Fig. A.1 become active in a series fashion. For example, in the 'StrongARM' comparator, the nMOS pair turns on first while the pMOS pair becomes active only after *Vout*/*Vout'* drop below $V_{dd} - V_T$. Therefore the effective total transconductance at the initial step of the regeneration phase is only half of that of the proposed comparator. A larger effective transconductance allows a faster regeneration speed and hence shorter delay time.

Simulation

The proposed comparator is designed in *TSMC* 65 nm CMOS technology. Its performance is simulated and compared with the other two dynamic comparators shown in Fig. A.1a, b which are also designed in the same technology. For a fair comparison, all three comparators in the testbench are designed to have a similar input referred offset voltage, and with the same external loading capacitance C_{ext} ($\sim$5 fF).

Fig. A.6 shows the simulated delay time (ns) of each comparator versus the supply voltages with an input voltage different of 50 mV. The delay time is defined as the time the output difference takes to reach $\frac{1}{2}V_{dd}$. The delay time is a suitable quantity in characterizing the comparator's speed of operation [7], a shorter delay time means a faster comparison speed. In Fig. A.6, it is shown that the delay time of the proposed comparator is about 30 % shorter than that of the 'Double-tail' comparator with 0.5 V supply voltage and it is only one third compared to that of the 'StrongARM' comparator with 0.6 V supply voltage.

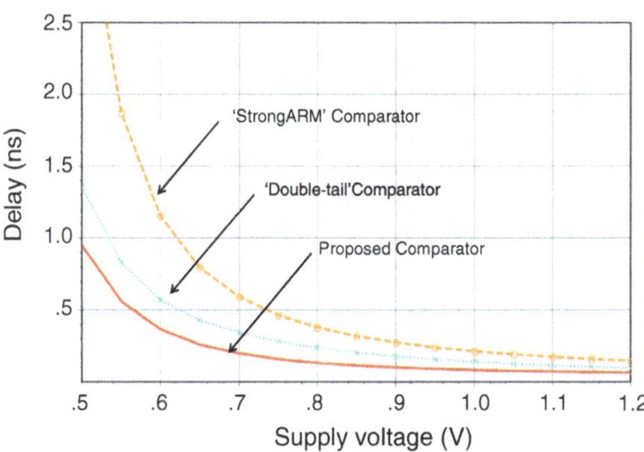

Fig. A.6 Delay time versus supply voltage (Δvin = 50 mV, $V_{cm} = V_{dd} - 0.1$ V)

Summary

A dynamic latch comparator suitable for very low supply voltage applications is presented in this appendix. Analysis and simulation show that it is able works at very lower supply voltage and with faster speed compared to the previous works (the 'StrongARM' and the 'Double-tail' latched comparators [3–4]). Moreover, it is a fully dynamic circuit without static power consumption.

References

1. McCarroll, B.J., C.G. Sodini, and H.-S. Lee. 1988. A high-speed CMOS comparator for use in an ADC. *IEEE Journal of Solid-State Circuits* 23(1): 159–165.
2. Y. Lin, K. Doris, H. Hegt, and A. van Roermund. 2012. A dynamic latched comparator for low supply voltages down to 0.45 V in 65-nm CMOS. In *2012 IEEE International Symposium on Circuits and Systems (ISCAS)*, pp. 2737–2740.
3. Kobayashi, T., K. Nogami, T. Shirotori, and Y. Fujimoto. 1993. A current-controlled latch sense amplifier and a static power-saving input buffer for low-power architecture. *IEEE Journal of Solid-State Circuits* 28(4): 523–527.
4. Schinkel, D., Mensink, E., Klumperink, E., Van Tuijl, E. and B. Nauta. 2007. A double-tail latch-type voltage sense amplifier with 18 ps Setup+Hold time. In *IEEE International Solid-State Circuits Conference Digest of Technical Papers(ISSCC 2007)*, pp. 314–605.
5. Razavi, B., K.F. Lee, and R.-H. Yan. 1995. Design of high-speed low-power frequency dividers and phase-locked loops in deep submicron CMOS. *IEEE Journal of Solid-State Circuits* 30: 101–109.
6. Kim, J., B.S. Leibowitz, J. Ren, and C.J. Madden. 2009. Simulation and analysis of random decision errors in clocked comparators. *IEEE Transactions on Circuits and Systems I: Regular Papers* 56(8): 1844–1857.
7. Yin, G.M., Op't Eynde, F. and W. Sansen. 1992. A high-speed CMOS comparator with 8-b resolution. *IEEE Journal of Solid-State Circuits*, 27(2): 208–211 (Feb. 1992).

Zeitfracht Medien GmbH
Ferdinand-Jühlke-Straße 7
99095 Erfurt, Deutschland
produktsicherheit@kolibri360.de